Lecture Notes in Mathematics

Edited by A. Dold, B. Eckmann and F. Takens

1457

O. Axelsson L. Yu. Kolotilina (Eds.)

Preconditioned Conjugate Gradient Methods

Proceedings of a Conference held in Nijmegen, The Netherlands, June 19–21, 1989

Springer-Verlag
Berlin Heidelberg New York London
Paris Tokyo Hong Kong Barcelona

Editors

Owe Axelsson
University of Nijmegen
Faculty of Mathematics and Informatics
Toernooiveld 1
6525 Nijmegen, The Netherlands

Lily Yu. Kolotilina
Steklov Mathematical Institute
LOMI
Fontanka 27, Leningrad D-111, USSR

Mathematics Subject Classification (1980): 65F10, 65N20, 65N30

ISBN 3-540-53515-2 Springer-Verlag Berlin Heidelberg New York
ISBN 0-387-53515-2 Springer-Verlag New York Berlin Heidelberg

Printing and binding: Druckhaus Beltz, Hemsbach/Bergstr.
2146/3140-543210 – Printed on acid-free paper

Preface

An International Conference on Preconditioned Conjugate Gradient Methods was held at the Faculty of Mathematics and Informatics, University of Nijmegen, The Netherlands on June 19-21, 1989.

The main motivation in organizing this Conference was the wish to bring together specialists working on iterative solution methods, in particular using preconditioning methods. The conference was preceeded by a short course on preconditioned conjugate gradient methods, on June 15-17, 1989.

Although the conference was organized and announced only a few months prior to its taking place, many scientists from different countries attended the conference. The director of the faculties of sciences, dr.ir. L.H.J. Wachters, had kindly accepted the invitation to open the conference.

The topics presented at the conference contained both analysis and implementational aspects of the methods. The proceedings contains the full text of 11 invited or contributed papers.

We are grateful to the participants, and especially to the speakers who made this meeting an important scientific event. We are also most indebted to the Administration of the Faculty of Mathematics and Informatics, to the younger members of the numerical analysis group and to the secretatiat of the Institute of Mathematics whose support was an indispensable contribution to the success of this conference.

O. Axelsson and L. Kolotilina

Table of Contents

MODIFIED INCOMPLETE FACTORIZATION STRATEGIES

R. BEAUWENS
UNIVERSITÉ LIBRE DE BRUXELLES, SERVICE DE MÉTROLOGIE NUCLÉAIRE
50, AV. F.D. ROOSEVELT, B-1050 BRUSSELS, BELGIUM

Abstract. We review here the conditioning properties of modified incomplete symmetric factorizations of Stieltjes matrices, appropriate for the PCG solution of symmetric positive definite linear systems. Emphasizing the algorithmic interpretation of the required assumptions, we analyse the theoretical support that can be given to practical factorization algorithms and the choice left open between "perturbed" and "unperturbed" policies. Recent results extending the scope of unperturbed modified factorizations are included and discussed.

1. Introduction. We consider in this work the PCG solution of a linear system

$$Ax = b \tag{1.1}$$

with the purpose of reviewing a priori upper bounds on the spectral condition number

$$\kappa(B^{-1}A) = \frac{\lambda_{max}(B^{-1}A)}{\lambda_{min}(B^{-1}A)} \tag{1.2}$$

of the preconditioned matrix $B^{-1}A$ and factorization "strategies" that are supported or suggested by these results, under the assumptions that A is a Stieltjes matrix and that B is determined from a modified incomplete factorization of A.

We notice here that the case of an arbitrary symmetric positive definite matrix A may be reduced to the Stieltjes case (or even to some more restrictive class of matrices) provided that one can determine spectral bounds for a pencil of the form $A - \lambda A_o$ where A_o is Stieltjes since, if α and β are positive numbers such that, for all $z \in C^n, z \neq 0$,

$$\alpha \leq \frac{(z, Az)}{(z, A_o z)} \leq \beta, \tag{1.3}$$

then

$$\kappa(B^{-1}A) \leq \frac{\beta}{\alpha}\kappa(B^{-1}A_o). \tag{1.4}$$

The assumption that B is determined from an incomplete factorization of A means that B is of the form

$$B = U^T P^{-1} U \tag{1.5}$$

where U is upper triangular with positive diagonal entries, $P = diag(U)$, and that the entries of $U = (u_{ij})$ are related to those of $A = (a_{ij})$ through the relations

$$u_{ij} = a_{ij} - \beta_{ij} \sum_{r<i} \frac{u_{ri}u_{rj}}{u_{rr}} \qquad i \leq j \tag{1.6}$$

where $\beta = (\beta_{ij})$ is a given (0,1) matrix. The assumption, made here, that B is determined from a *modified* incomplete factorization of A means that the determination of the diagonal entries of U is modified with respect to the previous scheme. These entries are now determined from the relations

$$B\hat{x} = A\hat{x} + \Lambda D\hat{x} \tag{1.7}$$

where $\hat{x}$ is a given positive vector, $D = diag(A)$ and Λ is a nonnegative diagonal matrix. The latter relations may equivalently be written

$$U\hat{x} = A\hat{x} + \Lambda D\hat{x} - (U^T - P)P^{-1}U\hat{x}. \tag{1.8}$$

In other words, modified incomplete factorizations use the relations (1.6) to determine $(U - P)$ and (1.8) to determine $U\hat{x}$, whence P.

The matrix $\Lambda = (\lambda_i \delta_{ij})$ is a nonnegative diagonal matrix of "small" parameters λ_i that may be given (static factorization algorithms) or that may be determined during the computation of U (dynamic factorization algorithms). The proper choice of Λ is crucial for the convergence behaviour of the associated PCG method and we shall review below the recipes that can be given for its choice on the basis of available upper bounds on $\kappa(B^{-1}A)$.

Standard definitions and notation used throughout the paper are the following. All matrices are $n \times n$ matrices and all vectors are n vectors. The order relation between real matrices and vectors is the usual componentwise order : with $A = (a_{ij})$ and $B = (b_{ij})$, then $A \leq B$ $(A < B)$ if $a_{ij} \leq b_{ij}$ $(a_{ij} < b_{ij})$ for all i, j. A is said nonnegative (positive) if $A \geq 0$ $(A > 0)$ and monotone if it is nonsingular with $A^{-1} \geq 0$. An M-matrix is a monotone matrix with nonpositive offdiagonal entries. A Stieltjes matrix is a symmetric M-matrix. If $A = (a_{ij})$, we denote by $diag(A)$ the (diagonal) matrix with entries $a_{ii}\delta_{ij}$. By e we denote the vector with all components equal to unity.

2. Spectral Bounds. We first review the most general spectral bounds obtained to date for the pencil $A - \nu B$, trying to emphasize the algorithmic interpretation of the assumptions required by these results. The following general assumptions will be made in most theorems reported in this work and referred to as (GA). Proofs that can be found elsewhere are omitted. See [5,6,8,10] for those of the results reported in the present section.

DEFINITION 2.1. *General assumptions (GA) mean the following.*

$$\begin{aligned}
&A \text{ is a real symmetric matrix},\\
&D = diag(A),\\
&U \text{ is a real upper triangular matrix},\\
&P = diag(U) \text{ is nonnegative and nonsingular},\\
&B = U^T P^{-1} U,\\
&\hat{x} \text{ is a given positive vector.}
\end{aligned}$$

We first separate upper and lower spectral bounds to display the proper relevance of additional assumptions.

THEOREM 2.2. *Adding to (GA) that*

$$U \text{ is an } M-matrix, \tag{2.1}$$

$$offdiag(U^T + U) \leq offdiag(A), \tag{2.2}$$

$$B\hat{x} \geq (1 - \tau)A\hat{x}, \tag{2.3}$$

$$U\hat{x} \geq (1 - \tau)P\hat{x}, \tag{2.4}$$

$$\tau < 1, \tag{2.5}$$

then

$$\lambda_{max}(B^{-1}A) \leq \frac{1}{1-\tau} \quad . \tag{2.6}$$

THEOREM 2.3. *Adding to (GA) that*

$$offdiag(A - (U^T - P)P^{-1}(U - P)) \leq offdiag(U^T + U), \tag{2.7}$$

$$B\hat{x} \leq A\hat{x}, \tag{2.8}$$

then

$$\lambda_{min}(B^{-1}A) \geq 1. \tag{2.9}$$

Gathering now these results, we get :

THEOREM 2.4. *Adding to (GA) that*

(2.10) U *is an* $M-matrix$,

(2.11) $offdiag(A - (U^T - P)P^{-1}(U - P)) \leq offdiag(U^T + U) \leq offdiag(A)$,

(2.12) $(1-\tau)A\hat{x} \leq B\hat{x} \leq A\hat{x}$,

(2.13) $U\hat{x} \geq (1-\tau)P\hat{x}$,

(2.14) $\tau < 1$,

then

$$1 \leq \lambda(B^{-1}A) \leq \frac{1}{1-\tau} \quad , \tag{2.15}$$

whence

$$\kappa(B^{-1}A) \leq \frac{1}{1-\tau} \quad . \tag{2.16}$$

The additional assumptions required by these results may be subdivided in two groups.

The first group, comprising (2.1), (2.2) and (2.7) or, in the combined version, (2.10) and (2.11), bears on the offdiagonal entries of U and A. They essentially determine the class of matrices A and B that are covered by the results. It is readily checked that, for incomplete or modified incomplete factorizations, (2.10) and (2.11) are always satisfied when A is a Stieltjes matrix. We refer to [10] for a recent generalization of these assumptions.

The second group, comprising (2.3), (2.4), (2.5) and (2.8) or, in the combined version (2.12), (2.13), (2.14), bears on the diagonal entries of U and deserves careful attention.

It should first be noticed that compatibility of (2.3) and (2.8) requires that $\hat{x}$ is chosen such as $A\hat{x} \geq 0$. This again is always possible when A is a Stieltjes matrix. The existence of $\hat{x} > 0$ with $A\hat{x} > 0$ is indeed a well-known criterion for A with $offdiag(A) \leq 0$ to be an M-matrix (cf. [18] : Theorem 1). In that case, $A\hat{x} > 0$ is further a sufficient condition for the existence of B satisfying (1.5)-(1.7). Existence criteria useful under the less stringent condition $A\hat{x} \geq 0$ will be mentioned below.

It should next be noticed that, if P is so chosen (i.e. large enough) that $U\hat{x} \geq \alpha P\hat{x}$ with $\alpha = 1 - \tau > 0$, then, by Theorem 2.2, $1/\alpha$ is an upper spectral bound for the pencil $A - \nu B$. For that purpose, it is sufficient to increase appropriate components of $B\hat{x}$ as it is readily seen from the relation

$$U\hat{x} = B\hat{x} - (U^T - P)P^{-1}U\hat{x}. \tag{2.17}$$

But $B\hat{x}$ may not be arbitrarily increased because of condition (2.8) of Theorem 2.3. The best compromise within the constraints (2.12) appears therefore to be

$$B\hat{x} = A\hat{x} \tag{2.18}$$

and, if this choice also meets the condition (2.13) with (2.14), then we get the bound (2.16). In other cases, one may try to increase $B\hat{x}$ beyond the limit $A\hat{x}$ (hereby requiring an improved version of Theorem 2.3) or philosophically live with $B\hat{x} = A\hat{x}$ (hoping that an improved version of Theorem 2.2 does apply).

Before proceeding to the consideration of these "strategies", we wish to figure out the order of magnitude of the bound $1/(1-\tau)$ with an analytical estimate which, although less accurate, is more appropriate for that purpose. It should however be appreciated that τ is readily computed during the factorization and thus, always numerically accessible.

Analytical results to be stated here and in later sections rest on the consideration of matrix graph properties and we now need a short digression into this area to recall some terminology. We refer to [11] or [14] for the general terminology on matrix graphs with the warning that all graphs considered in this work are ordered undirected graphs with node set $[1,n]$, i.e. the ordered set of the first n integers or, when subgraphs are considered, some subset of $[1,n]$. Given a graph G, we use the notation $Adj(i)$ to denote the set of neighbors of i in G; with $M \subset [1,n]$, we further set

$$Adj(M) = \bigcup_{i \in M} Adj(i).$$

DEFINITION 2.5. *An increasing path in a graph is a path $i_o, i_1, i_2, \ldots i_\ell$ such that $i_o < i_1 < i_2 < \ldots < i_\ell$.*

DEFINITION 2.6. *The maximal increasing length $\ell(M)$ of an nonempty subset M of the node set of the graph G is the length of the longest increasing path in the subgraph of G induced by M. We further set $\ell(\emptyset) = -1$.*

DEFINITION 2.7. *A node k of a graph G is called a precursor (successor) of the node i of G if $k \in Adj(i)$ with $k < i (k > i)$. The set of precursors (successors) of i is denoted by $P(i)$ $(S(i))$. If M is a subset of the node set of G, the set of precursors (successors) of the nodes of M is denoted by $P(M)$ $(S(M))$. In other words :*

$$P(M) = \bigcup_{i \in M} P(i), \qquad S(M) = \bigcup_{i \in M} S(i).$$

We further set $P(\emptyset) = S(\emptyset) = \emptyset$.

DEFINITION 2.8. *For any node i of a graph G, we define the ascent $As(i)$ of i as*

$$As(i) = \{k \mid \text{ There exists an increasing path from } k \text{ to } i\}.$$

For any set M of nodes of G, we define the ascent As(M) of M as

$$As(M) = \bigcup_{i \in M} As(i)$$

if $M \neq \emptyset$, *with* $As(\emptyset) = \emptyset$.

DEFINITION 2.9. *For any pair of nodes i and j of a graph G, we denote by*

$$Pc(i,j) = P(i) \cap P(j)$$

their set of common precursors. We further set

$$Pc(G) = \bigcup_{\substack{i,j=1 \\ i \neq j}}^{n} Pc(i,j)$$

and, if G is the graph of a matrix A, we also use Pc(A) for Pc(G).

It may be mentioned as a first application of these definitions, that the assumption (2.4) of Theorem 2.2 (whence also (2.13) of Theorem 2.4) may be weakened to

$$(U\hat{x})_i \geq (1-\tau)(P\hat{x})_i \qquad for \qquad i \in Pc(U) \quad , \tag{2.19}$$

a remark which leads us to the following analytical bound.

THEOREM 2.10. *Under (GA),* (2.1), (2.2) *and*

$$\frac{1}{k+\ell_i+1}((U^T-U)\hat{x})_i \leq (B\hat{x})_i \qquad for \qquad i \in As(Pc(U)) \tag{2.20}$$

$$\frac{1}{k+\ell+2} A\hat{x} \leq B\hat{x} \tag{2.21}$$

with $\ell_i = \ell(As(i))$, $\ell = \ell(As(Pc(U)))$ *and* $k \geq 0$, *we have that the conditions* (2.3) *and* (2.19) *are satisfied with* $1-\tau = 1/(k+\ell+2)$. *Therefore*

$$\lambda_{max}(B^{-1}A) \leq k+\ell+2 \tag{2.22}$$

and, if (2.7) *and* (2.8) *are also met,*

$$\kappa(B^{-1}A) \leq k+\ell+2. \tag{2.23}$$

This result displays a relation between the ordering of U and the condition number of $B^{-1}A$ since the assumption (2.20) and the conclusion (2.22) (or (2.23)) clearly depend on the ordering of U. One may recommend, on this basis, to order U so as to satisfy (2.20) (together with $B\hat{x} = A\hat{x}$ as already discussed) with the smallest possible value of $k+\ell$.

Consideration of applications to discrete *PDE*'s shows on the other hand that this bound may be considered as an algebraic generalization of the $O(h^{-1})$ bound of the geometrical approach developed by Axelsson and his co-workers (cf. [2,13] and the references cited there).

It must finally be remarked that the conditions (2.19) and (2.14) require the strict diagonal dominance of U with respect to $Pc(U)$ and $\hat{x}$ according to the following definition.

DEFINITION 2.11. *Let $A = (a_{ij})$ be a matrix with positive diagonal entries and non-positive offdiagonal entries, let x be a positive vector and $M \subset [1,n]$. We call dominance ratio of A with respect to M and x the number*

$$t = \max_{i \in M} \left(\frac{-\sum_{j \neq i} a_{ij} x_j}{a_{ii} x_i} \right)$$

with $t = 0$ if $M = \emptyset$. A is diagonally dominant (with respect to M and x) if $t \leq 1$ and strictly diagonally dominant (with respect to M and x) if $t < 1$.

Before concluding this section,let us reconsider the existence problem of B subject to (1.5)-(1.7) when A is an M-matrix with $\hat{x} > 0$ such that $A\hat{x} \geq 0$. Notice first that Gustafsson's existence analysis (cf. [13] : Theorem 3.1) precludes the unperturbed case by requiring a strictly positive perturbation $\Lambda D\hat{x}$. The latter condition acts like strict diagonal dominance (i.e. $A\hat{x} > 0$) and dispenses with any additional requirement. Notice on the other hand that the author's criterion (cf. [5] : Corollary of Theorem 2.1) of lower semistrict diagonal dominance (i.e. $A\hat{x} \geq 0$ with $\sum_{j=1}^{j=i} a_{ij} x_j > 0$) is sufficient but not necessary. The precise necessary and sufficient additional condition has recently been obtained by Notay (cf. [17] : Theorem 3.4) and writes in the present framework, with graph notation referring to $G(U)$,

$$\forall j : S(j) = \emptyset \Rightarrow \exists i \in As(j) \textit{ with } (A\hat{x})_i + \lambda_i (D\hat{x})_i > 0 \tag{2.24}$$

and in particular, when $\Lambda = 0$,

$$\forall j : S(j) = \emptyset \Rightarrow \exists i \in As(j) \textit{ with } (A\hat{x})_i > 0. \tag{2.25}$$

It may further be added that (2.25) is true a.o. whenever the only nodes without successors are the last nodes of the connected components of $G(U)$, a condition which occurs under the assumptions considered in Section 4.

3. Factorizations algorithms. Let us now recall that modified incomplete factorizations use the relations (1.6) to determine $(U - P)$ and (1.8) to determine $U\hat{x}$, whence P. We consider in this section that A is a Stieltjes matrix and that β is a given (0,1) matrix. It then follows from (1.6) that the assumptions (2.1), (2.2) and (2.7) of Theorems 2.2 and 2.3 are satisfied.

It remains to determine P. We consider here four examples of algorithms that may be used for that purpose following the "strategies" indicated in the previous section and we try to analyse their properties on the basis of the preceding theory.In all cases, a (small) parameter α with $0 < \alpha < 1$ and a positive vector $\hat{x}$ have to be chosen; $1/\alpha$ is the "target" upper bound on $\kappa(B^{-1}A)$; $\hat{x}$ is used to measure the degree of diagonal dominance of U and must be such that $A\hat{x} \geq 0$. We also assume here that (2.25) is satisfied. A simple modification dispensing us from that requirement will be indicated in fine.

It is understood that each line of U is computed successively beginning with u_{ij} for $i < j \leq n$ (cf. (1.6)) and ending with u_{ii} (cf. below). It is clear that, at the ith stage, the formula (1.6) needs u_{ss} for $s < i$ only.

STRATEGY 1. *Compute $P = diag(U)$ by solving, at the ith stage,*

$$(U\hat{x})_i = (A\hat{x})_i - ((U^T - P)P^{-1}U\hat{x})_i \tag{3.1}$$

for u_{ii}.

STRATEGY 2. *Denote by $P_o^{(i)}$ the diagonal matrix equal to $P = diag(U)$ with the (possible) exception of its ith (diagonal) entry which we denote by u_{ii}^o and let $U_o^{(i)} = P_o^{(i)} + (U - P)$. Notice that neither $P_o^{(i)}$ nor P is entirely known at the ith stage but this is irrelevant for our purpose.*

Determine first u_{ii}^o by solving

$$(U_o^{(i)}\hat{x})_i = (A\hat{x})_i - ((U - P)^T P^{-1} U\hat{x})_i \tag{3.2}$$

for u_{ii}^o and compute $\alpha_o^i = (U_o^{(i)}\hat{x})_i/(P_o^{(i)}\hat{x})_i$. If $i \notin As(Pc(U))$ or if $\alpha_o^i \geq \alpha$, set $u_{ii} = u_{ii}^o$. Otherwise, determine u_{ii} by solving

$$(U\hat{x})_i = (A\hat{x})_i + \delta_i(P_o^{(i)}\hat{x})_i - ((U - P)^T P^{-1} U\hat{x})_i \tag{3.3}$$

for u_{ii}, where δ_i is given by

$$\delta_i = \frac{\alpha^2}{1-\alpha} + \frac{\alpha}{1-\alpha} \frac{max\{((U^T - U)\hat{x})_i, 0\}}{(P_o^{(i)}\hat{x})_i}. \tag{3.4}$$

STRATEGY 3. *Proceed in the same way as in the former case, but define δ_i by*

$$\delta_i = \frac{\alpha^2}{1-\alpha} + \frac{\alpha}{1-\alpha} \frac{((U^T - U)\hat{x})_i}{(P_o^{(i)}\hat{x})_i}. \tag{3.5}$$

STRATEGY 4. *Determine u_{ii}^o, $P_o^{(i)}$, $U_o^{(i)}$ and α_o^i as in the former case. If $i \notin Pc(U)$ or if $\alpha_o^i \geq \alpha$, set $u_{ii} = u_{ii}^o$. Otherwise, determine u_{ii} by solving (3.3) for u_{ii} with*

$$\delta_i = \frac{\alpha - \alpha_o^i}{1-\alpha}. \tag{3.6}$$

The first strategy was called "unmodified" factorization in [2] (with $\hat{x} = e$). We prefer to call it "unperturbed" to avoid confusion with the case where P is also determined by the relation (1.6) and because it amounts to set $\Lambda = 0$ in the relation (1.7). The second strategy is an obvious adaptation to our framework of the Axelsson-Barker scheme (defined by Eqs (7.18) in [2]) and the third one is a simplified version of the same algorithm. The fourth strategy is new although supported by the same philosophy as we shall see below. Notice that in all cases, the diagonal matrix $\Lambda = (\lambda_i\delta_{ij})$ of Eq. (1.7) may be determined from

$$\lambda_i = \frac{u_{ii}^o}{a_{ii}} \delta_i \tag{3.7}$$

provided that δ_i is initialized to zero at the beginning of the ith stage.

The first remark to be done when discussing these strategies is that, would it happen that $\alpha_o^i \geq \alpha$ at all stages, then all four strategies reduce to the first one. It does happen in particular when the assumptions of Theorem 2.10 are satisfied, with $\alpha = \frac{1}{k+\ell+2}$ and this shows that the indications provided by the latter result should be used before starting any of these procedures to :

(1) try to find an ordering of U such that $k+\ell$ be not too large and that the condition (2.20) is approximately satisfied at most if not all nodes of $As(Pc(U))$;

(2) choose α once the ordering has been chosen.

Our next remark is to emphasize the opposite points of view which support the first strategy on the one hand and the other three on the other hand : the first strategy preserves an exact lower bound for the pencil $A - \nu B$ since 1 is then the lowest eigenvalue of $A - \nu B$ (with $\hat{x}$ as associated eigenvector) while, as it will now be shown, the other strategies imply that $1/\alpha$ is an upper bound on the spectrum of $A - \nu B$.

THEOREM 3.1. *Adding to (GA) that*

(3.8) $\quad A$ *is a Stieltjes matrix with* $A\hat{x} \geq 0$,

(3.9) $\quad U - P$ *satisfies* (1.6) *with* $0 \leq \beta_{ij} \leq 1$ *for* $i < j \leq n$,

(3.10) $\quad P$ *is determined from strategy n°i with* $i = 2$ *or* 3 *or* 4 ,

(3.11) $\quad 0 < \alpha < 1$,

then (2.3) *and* (2.19) *are satisfied with* $1 - \tau = \alpha$. *Therefore*

$$\lambda_{max}(B^{-1}A) \leq \frac{1}{\alpha} \quad . \tag{3.12}$$

Proof. It is clearly sufficient to show that the ith stage leads to u_{ii} such that $(U\hat{x})_i \geq \alpha(P\hat{x})_i$, whenever $i \notin Pc(U)$.

In the case of the fourth strategy, this follows from

$$\begin{aligned}(U\hat{x})_i - \alpha(P\hat{x})_i &= (U_o^{(i)}\hat{x})_i + \delta_i(P_o^{(i)}\hat{x})_i - \alpha((P_o^{(i)}\hat{x})_i + \delta_i(P_o^{(i)}\hat{x})_i) \\ &= (\alpha_o^i - \alpha + \delta_i(1-\alpha))(P_o^{(i)}\hat{x})_i = 0\end{aligned}$$

when $\alpha_o^i < \alpha$ while there is nothing to prove when $\alpha_o^i \geq \alpha$.

In the case of the third strategy, we prove by induction that $(U\hat{x})_i \geq \alpha(P\hat{x})_i$ whenever $i \in As(Pc(U))$ (whence also in particular for $i \in Pc(U)$). If $\alpha_o^i < \alpha$, we have

$$\begin{aligned}(U\hat{x})_i - \alpha(P\hat{x})_i &= (U_o^{(i)}\hat{x})_i + ((1-\alpha)\delta_i - \alpha)(P_o^{(i)}\hat{x})_i \\ &= (U_o^{(i)}\hat{x})_i + \alpha((U^T - U)\hat{x})_i - \alpha(1-\alpha)(P_o^{(i)}\hat{x})_i \\ &= (U_o^{(i)}\hat{x})_i + \alpha(U^T - P)\hat{x})_i \\ &\qquad - \alpha((U-P)\hat{x})_i - \alpha(1-\alpha)(P_o^{(i)}\hat{x})_i \\ &= (A\hat{x})_i - ((U^T - P)P^{-1}(U - \alpha P)\hat{x})_i \\ &\qquad + \alpha(1-\alpha_o^i)(P_o^{(i)}\hat{x})_i - \alpha(1-\alpha)(P_o^{(i)}\hat{x})_i \\ &= (A\hat{x})_i - ((U^T - P)^{-1}(U - \alpha P)\hat{x}_i + \alpha(\alpha - \alpha_o^i)(P_o^{(i)}\hat{x})_i \quad ,\end{aligned}$$

but the first term of the RHS of the latter equation is nonnegative and the last one positive by assumption while the second one is nonnegative by the induction hypothesis, entailing

$$(U\hat{x})_i - \alpha(P\hat{x})_i > 0$$

while again, there is nothing to prove if $\alpha_o^i \geq \alpha$.

The same argument holds a fortiori in the case of the second strategy. ■

It is now clear that the last three strategies satisfy all the assumptions of Theorem 2.4 with the possible exception of the right inequality (2.12). Defining $\Lambda = (\lambda_i\delta_{ij})$ through (3.7) we actually have

$$B\hat{x} = A\hat{x} + \Lambda D\hat{x}$$

and the question raises to evaluate the influence of the perturbation $\Lambda D\hat{x}$ on the lower spectral bound of $A - \nu B$. One has the following result.

THEOREM 3.2. *Under* (*GA*), (2.7) *and*

$$B\hat{x} \leq A\hat{x} + \Lambda D\hat{x} \tag{3.13}$$

$$\Lambda = (\lambda_i \delta_{ij}) \geq 0 \tag{3.14}$$

$$\xi = \max_{z \neq 0} \frac{(z, \Lambda D z)}{(z, Az)} \tag{3.15}$$

one has

$$\lambda_{min}(B^{-1}A) \geq \frac{1}{1+\xi} \quad . \tag{3.16}$$

Proof. We have by (2.7) that

$$offdiag(A + \Lambda D - B) = offdiag(A - B) \leq 0$$

and, by (3.13) with (3.14) that

$$(A + \Lambda D - B)\hat{x} \geq 0.$$

It follows that $A + \Lambda D - B$ is nonnegative definite, whence by (3.15),

$$(z, Bz) \leq (z, (A + \Lambda D)z) \leq (z, Az)(1 + \xi)$$

which implies (3.16). ∎

This result reduces the question raised above to the estimation of ξ. Clearly

$$\xi \leq \frac{\lambda}{\lambda_1(D^{-1}A)} \quad , \tag{3.17}$$

where

$$\lambda = \max_{1 \leq i \leq n} (\lambda_i) \tag{3.18}$$

and

$$\lambda_1(D^{-1}A) = \min_{z \neq 0} \frac{(z, Az)}{(z, Dz)} \quad . \tag{3.19}$$

But this bound on ξ is often too large; more precisely, with

$$< \lambda > = \frac{(\hat{x}, \Lambda D\hat{x})}{(\hat{x}, D\hat{x})} \quad , \tag{3.20}$$

it is found that (3.17) is accurate when $\lambda \simeq < \lambda >$ but an order of magnitude too large when $\lambda \gg < \lambda >$ (see [6] and [8] for a more detailed account).

Unfortunately, no closed form formula has been rigorously obtained to deal with the latter case. Instead, appropriate procedures have been set up for obtaining upper bounds on ξ in specific situations. Since these procedures are somewhat involved, we must however refer the reader to the literature (cf. [2,8,12,15]) for more detailed information.

Nevertheless, the following heuristic formula can be proposed

(3.21) $$\xi \simeq \frac{<\lambda>}{\lambda_1(D^{-1}A)}$$

inasmuch as one may admit that the fundamental eigenvector z_1 of th pencil $A - \lambda D$ is a valid trial vector for the Raleigh quotient estimate of ξ, i.e. that

(3.22) $$\xi \simeq \frac{(z_1, \Lambda D z_1)}{(z_1, A z_1)} = \frac{(z_1, \Lambda D z_1)}{(z_1, D z_1)} \frac{1}{\lambda_1(D^{-1}A)}$$

and further that

(3.23) $$\frac{(z_1, \Lambda D z_1)}{(z_1, D z_1)} \simeq <\lambda> .$$

Despite the lack of rigorous justification of the formula (3.21), it has the essential merit of representing the most concise summary of the general conclusions that can be drawn from the rigorous analysis of a variety of examples to be found a.o. in [2,8,12,15], reproducing the exact formula when $\lambda = <\lambda>$, exhibiting the right order of magnitude when $\lambda \gg <\lambda>$ and displaying correctly the influence of $\lambda_1(D^{-1}A)$.

The main practical conclusion to be drawn from the latter remark is that the strategies N° 2, 3 and 4 are likely to provide a good spectral condition number $\kappa(B^{-1}A)$ whenever $<\lambda> \simeq \lambda_1(D^{-1}A)$, i.e. whenever $\lambda_1(D^{-1}A)$ is not too small. In other cases (quasi singular problems) one may attempt to limit the perturbations λ_i so that $<\lambda> \simeq \lambda_1(D^{-1}A)$, i.e. essentially shift to the first strategy. As noticed in Section 2, such an attitude relies on the hope that an improved version of Theorem 2.2 applies. The results to be reported in the next section show that such an hope needs not be unrealistic.

It is now appropriate to reconsider the assumption (2.25). It follows from the (2.24) version of Notay's result that a simple way to take care of any violation of (2.25) at some node j is to introduce a corresponding positive perturbation $\lambda_j(D\hat{x})_j$. Since $S(j) = \emptyset$, this perturbation does not affect later coefficients nor does it affect the upper spectral bound discussed above. Since on the other hand, its size is arbitrary, it can be chosen small enough to have a negligible influence on the lower spectral bound. With this slight modification of our factorization algorithms, the only requirement to be put on $\hat{x}$ is that $A\hat{x} \geq 0$.

Before leaving this section, let us finally notice that the essential limitation of Theorem 2.2 arises from the requirement (2.5) which (together with (2.19)) means that U must be strictly diagonally dominant (with respect to $Pc(U)$ and $\hat{x}$). In this respect, it is of interest to mention the following result by Axelsson [1] (cf. also [3]) which applies to the generalized SSOR method.

THEOREM 3.3. *Adding to (GA) that :*

$$\begin{aligned} &A \text{ is symmetric positive definite,} \\ &offdiag(U^T + U) = offdiag(A), \\ &D\hat{x} \leq 2\,\tau_a\, P\hat{x}, \\ &\tau_a < 1, \end{aligned}$$

then

$$\lambda_{max}(B^{-1}A) \leq \frac{1}{2}\frac{1}{1-\tau_a} .$$

Exchanging Theorem 2.2 for the latter result, a new family of strategies can be developed (for the generalized SSOR factorization of Stieltjes matrices) in much the same way as done above, but where the perturbations have now to take care of the condition $\tau_a < 1$ (i.e. of the strict positive definiteness of $2P - D$) rather than $\tau < 1$. Unfortunately this new requirement is almost as restrictive as the preceding one and leads us to similar difficulties.

4. The unperturbed strategy. We now report recent results on upper spectral bounds which do neither require the strict diagonal dominance of U nor the strict positive definiteness of $2P - D$. These improved versions of Theorem 2.2 have been obtained in [9] and [16] under the additional assumption that the approximate triangular factors are "S/P consistently ordered". We therefore begin with a few definitions to clarify this notion.

We first rephrase Young's definition of consistent ordering [19,20] through the notion of level structure [11].

DEFINITION 4.1. *A level structure of a graph G is a partitioning of its node set, $\mathcal{L} = (L_k)_{k=0,1,\dots,\ell}$, such that*

$$\begin{array}{lll} (a) & Adj(L_o) \subset L_o \cup L_1 \quad ; & \\ (b) & Adj(L_k) \subset L_{k-1} \cup L_k \cup L_{k+1} & for \quad 0 < k < \ell \ ; \\ (c) & Adj(L_\ell) \subset L_{\ell-1} \cup L_\ell \quad ; & \end{array}$$

ℓ is called the length of $\mathcal{L}$.

DEFINITION 4.2. *A graph G is consistently ordered if there exists a level structure $\mathcal{L} = (L_k)_{k=0,1,\dots,\ell}$ of G such that*

$$i \in L_k \Rightarrow P(i) \subset L_{k-1} \quad for \quad 0 < k \le \ell$$

and

$$i \in L_k \Rightarrow S(i) \subset L_{k+1} \quad for \quad 0 \le k < \ell .$$

A matrix is said to be consistently ordered when its graph is consistently ordered.

A level structure satisfying the preceding properties will be called associated with the consistent ordering of G. As observed in [7], if G is consistently ordered and connected, it has a unique level structure associated with its consistent ordering.

Next, complementing Definition 2.9 with the notation $Sc(i,j)$ to denote the common successors of any pair of nodes i and j in a graph G :

$$Sc(i,j) = S(i) \cap S(j),$$

we introduce the notion of S/P consistent ordering.

DEFINITION 4.3. *We say that a graph G is S/P consistently ordered if, for any pair of nodes i, j with $i \neq j$,*

$$Pc(i,j) \neq \emptyset \Rightarrow Sc(i,j) \neq \emptyset \quad .$$

A matrix is said to be S/P consistently ordered when its graph is S/P consistently ordered.

We recall from [7] that an S/P consistently ordered graph is consistently ordered and that if it is connected, its (unique) associated level structure satisfies :

$$\begin{array}{ll} (a) & \textit{its last level is } L_\ell = \{n\} \quad ; \\ (b) & \textit{its } k^{th} \textit{ level is } L_k = P(L_{k+1}) \textit{ for } 0 \le k < \ell. \end{array}$$

To deal with S/P consistently ordered graphs and matrices, it is appropriate to further complement Definition 2.9 with the following notation : for any couple of nodes i and j in a graph G, we set

$$Sp(i,j) = \begin{cases} Sc(i,j) & \text{if} \quad Pc(i,j) \neq \emptyset \\ \emptyset & \text{otherwise} \end{cases}$$

and we define

$$Sp(G) = \bigcup_{\substack{i,j=1 \\ i \neq j}}^{n} Sp(i,j).$$

If G is the graph of the matrix A, we also write $Sp(A)$ for $Sp(G)$.

To introduce our next definition, let $U = (u_{ij})$ be an upper triangular matrix and consider products of the form $u_{ri}u_{rj}$ with $r < i < j$; such expressions are nonzero only if $r \in Pc(U)$; now if U is S/P consistently ordered, to nonzero expressions of this type, we can associate nonzero products of the form $u_{is}u_{js}$ with $i < j < s$ since the existence of $r \in Pc(i,j)$ entails that of $s \in Sc(i,j)$; further, if U is an M-matrix, we can introduce positive normalization constants u_{ijrs} such that $u_{is}u_{js}/u_{ijrs} = u_{ri}u_{rj}/u_{rr}$; more generally, to any pair of nodes i,j with $i \neq j$ such that $Pc(i,j) \neq 0$, we can associate some family of positive parameters u_{ijs}, $s \in Sc(i,j)$ such that

$$\sum_{s \in Sc(i,j)} \frac{u_{is}u_{js}}{u_{ijs}} = \sum_{r \in Pc(i,j)} \frac{u_{ri}u_{rj}}{u_{rr}}.$$

These remarks show that any upper triangular S/P consistently ordered M-matrix has S/P images according to the following definitions.

DEFINITION 4.4. *Let $U = (u_{ij})$ be an upper triangular matrix with positive diagonal entries. Then a lower triangular matrix $L = (\ell_{ij})$ is called an S/P image of U if*

(a) the offdiagonal entries of L satisfy

$$\ell_{ij} = \begin{cases} u_{ji} & \textit{if} \quad i \in Sp(U) \textit{ and } j \in S(Pc(U)), \\ 0 & \textit{otherwise,} \end{cases}$$

(b) the diagonal entries of L satisfy

$$\ell_{ss} = \begin{cases} \min\{u_{ijs} \,|\, i,j \in P(s), \quad i \neq j, \quad Pc(i,j) \neq \emptyset\} & \textit{if } s \in Sp(U), \\ 1 & \textit{otherwise,} \end{cases}$$

where u_{ijs} is a family of positive parameters defined for all $s \in Sp(U)$, $i,j \in P(s)$, $i \neq j$ and $Pc(i,j) \neq \emptyset$, and which satisfy the relations

$$\sum_{s \in Sc(i,j)} \frac{u_{is}u_{js}}{u_{ijs}} = \sum_{r \in Pc(i,j)} \frac{u_{ri}u_{rj}}{u_{rr}}$$

for all couples of indices i,j with $i \neq j$ and $Pc(i,j) \neq \emptyset$.

DEFINITION 4.5. *Let $U = (u_{ij})$ be an upper triangular S/P consistently ordered M-matrix and let the diagonal matrix $K = (\kappa_i \delta_{ij})$ be such that $0 < \kappa_i \leq 1$ for all $i \leq i \leq n$. Any S/P image of KU is called a reduced S/P image of U, of reduction matrix K and of reduction ratio $\kappa = \min_{1\leq i\leq n}(\kappa_i)$.*

With these notions in mind, we can now state the following improved version of Theorem 2.2, borrowed from [9].

THEOREM 4.6. *Under (GA), (2.1), (2.2), (2.19) and*

$$U \text{ is } S/P \text{ consistently ordered,} \tag{4.1}$$

$$L \text{ is a reduced } S/P \text{ image of } U, \text{ of reduction ratio } \eta, \tag{4.2}$$

$$B\hat{x} \geq (1 - \tau + \eta\,\frac{\tau}{\ell + 2})A\hat{x}, \tag{4.3}$$

$$(L\hat{x})_i \geq (1 - \tau)(Q\hat{x})_i \quad \text{for} \quad i \in Sp(U), \tag{4.4}$$

$$\tau \leq 1, \tag{4.5}$$

with $Q = diag(L)$ and $\ell = \ell(Pc(U))$, we have that

$$\lambda_{max}(B^{-1}A) \leq \frac{1}{1 - \tau + \eta\,\frac{\tau}{\ell+2}} \leq \frac{\ell + 2}{\eta} \tag{4.6}$$

and therefore, if (2.7) and (2.8) are also met,

$$\kappa(B^{-1}A) \leq \frac{1}{1 - \tau + \eta\,\frac{\tau}{\ell+2}} \tag{4.7}$$

The condition (4.5) together with (2.19) means that U is diagonally dominant with respect to $Pc(U)$ and $\hat{x}$ but strict diagonal dominance is no more required. We refer to [9] for numerical results illustrating the power of Theorem 4.6, noticing here that, when U is S/P consistently ordered, it justifies the recourse to the first strategy suggested in Section 3, at least when the other ones are not satisfactory.

It remains of course to consider the implications of the other assumptions of this new bound to be sure that it truly generalizes Theorem 2.2.

It should first be noticed in this respect that (4.2), (4.4) and (4.5) require the existence of a reduced S/P image of U which must be diagonally dominant with respect to $Sp(U)$ and $\hat{x}$. That this requirement is always satisfied has recently been shown by Notay who proved in [16] that any S/P consistently ordered upper triangular M-matrix has reduced S/P images of any a priori prescribed dominance ratio t (with respect to $Sp(U)$ and any given $\hat{x} > 0$).

Because of Notay's theorem, we may introduce the following definition.

DEFINITION 4.7. *Let U be an upper triangular S/P consistently ordered M-matrix, $\hat{x}$ a positive vector and t a positive number. We call maximal reduction ratio of U with respect to $\hat{x}$ and t and we denote by $\eta_t(U, \hat{x})$ the maximal value of η such that U has a reduced S/P image of reduction ratio η and whose dominance ratio with respect to Sp(U) and $\hat{x}$ does not exceed t.*

It follows from these considerations that the assumptions (4.2) and (4.4) may simply be exchanged for

$$\eta \leq \eta_\tau(U, \hat{x}). \tag{4.8}$$

On the other hand, when considering modified incomplete factorizations of a Stieltjes matrix A with $A\hat{x} \geq 0$ and $B\hat{x} = A\hat{x}$, the condition (4.3) is actually implied by $B\hat{x} = A\hat{x}$. The only restrictive assumption of Theorem 4.6 is thus (4.1) which will now be discussed.

The "S/P consistency" condition being quite recent, no rigorous analysis of its exact bearing is yet available and we are forced to limit our discussion to a certain number of conjectures suggested by our first observations.

We shall in this respect restrict ourselves to the consideration of 2-dimensional "rectangular" graphs (typical of the generalized SSOR preconditioning of the standard five-point finite difference approximation of 2-dimensional second order partial differential equations). Our feelings are that the present definition of "S/P consistency" is appropriate for d-dimensional rectangular graphs, that the discussion of the 2-dimensional case is sufficient for illustration and, on the other hand, that it is the definition itself which needs to be generalized to cover more general cases. From a practical point of view, we expect that all qualitative aspects of our discussion do have a much wider field of application.

For practical applications in the present status of the theory, one may at any rate use the reduction technique mentioned in the introduction to reduce the given matrix A to some Stieltjes matrix A_o with a rectangular graph, a technique which covers a wide class of applications as illustrated by the work of Axelsson and Gustafsson [4].

Now, considering rectangular graphs, it should first be noticed that no ordering of a rectangular graph with a hole (cf. G_d on Figure 1) can be S/P consistent because the smallest node of the hole boundary is a common precursor of its two neighbors which have by assumption no common successor. On the other hand, it was recently noticed by Notay[1] that the following algorithm produces an S/P consistent ordering on any rectangular mesh without holes : number any node as last node, number next unnumbered neighbors as next precursors. Figure 1 shows a few examples of rectangular graphs ordered by Notay's algorithm. It is readily seen that G_a, G_b and G_c are S/P consistently ordered. By comparison, while any lexicographic ordering of G_a is S/P consistent, it need not be such in the case of G_b and cannot in that of G_c. Further, Notay's ordering is particularly attractive when a central node is chosen as last node since this choice reduces the value of our basic parameter $\ell + 2$ which characterizes the order of magnitude of our upper spectral bound of $A - \nu B$.

Before concluding this discussion, we have to stress that the distinction between zero and nonzero entries, the only information transmitted from a matrix U to its graph $G(U)$, need not be sharp from a numerical point of view and, therefore, that an S/P consistently ordered matrix may sometimes behave as "nearly" not S/P consistently ordered. It has been suggested by Notay [16] to take care of this consideration by using $\eta_1(U, \hat{x})$ as a measure of the "degree" of S/P consistency of an upper triangular $M - matrix$ U (with $\eta_1(U, \hat{x}) = 0$ when U is not S/P consistently ordered). Following this suggestion, one may say that U is "sufficiently" S/P consistent when $\eta_1(U, \hat{x})$ is not too small and conclude that Theorems 2.4 and 4.6 qualify the unperturbed strategy whenever U is sufficiently diagonally dominant with respect to $Pc(U)$ and $\hat{x}$ or U is (only) diagonally dominant with respect to $Pc(U)$ and $\hat{x}$ but sufficiently S/P consistent.

REFERENCES

[1] O Axelsson, *On iterative solution of elliptic difference equations on a mesh- connected array of processors*, Intern. Journal of High Speed Computing, 1 (1989), pp. 165–184.

[1] Private communication

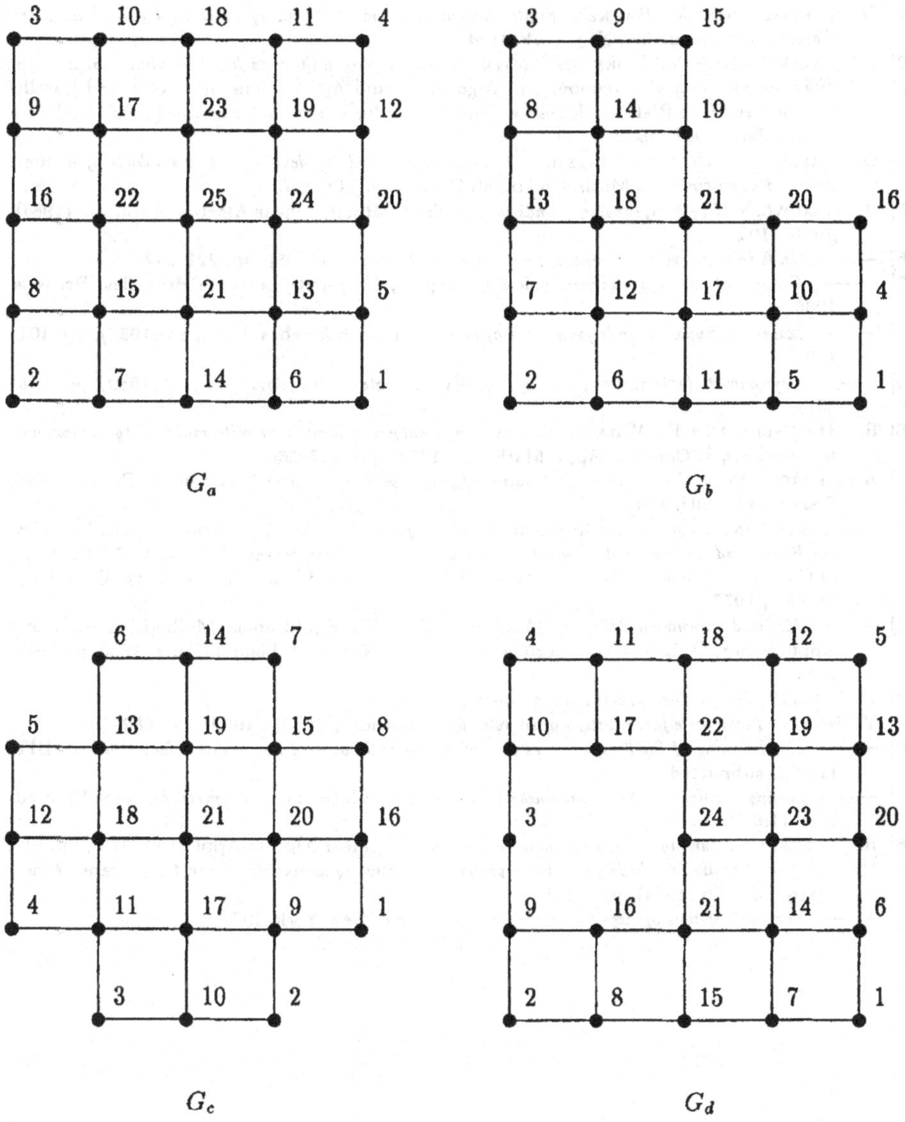

FIG. 1. *Notay's ordering for a few rectangular meshes*

[2] O. Axelsson and V. Barker, *Finite Element Solution of Boundary Value Problems. Theory and Computation*, Academic, New York, 1984.

[3] O. Axelsson and V. Eijkhout, *Robust vectorizable preconditioners for three-dimensional elliptic difference equations with anisotropy*, in Algorithms and Applications on Vector and Parallel Computers, H. te Riele, T. J. Dekker, and H. van der Vorst, eds., Elsevier (North Holland), Amsterdam, 1987, pp. 279–306.

[4] O. Axelsson and I. Gustafsson, *Preconditioning and two-level multigrid methods of arbitrary degree of approximation*, Math. Comp., 40 (1983), pp. 214–242.

[5] R. Beauwens, *Upper eigenvalue bounds for pencils of matrices*, Linear Algebra Appl., 62 (1984), pp. 87–104.

[6] ———, *On Axelsson's perturbations*, Linear Algebra Appl., 68 (1985), pp. 221–242.

[7] ———, *Consistent ordering analysis*, Scientific Report, Université Libre de Bruxelles, Brussels, 1986.

[8] ———, *Lower eigenvalue bounds for pencils of matrices*, Linear Algebra Appl., 85 (1987), pp. 101–119.

[9] ———, *Approximate factorizations with S/P consistently ordered M-factors*, BIT, 29 (1989), pp. 658–681.

[10] R. Beauwens and R. Wilmet, *Conditioning analysis of positive definite matrices by approximate factorizations*, J. Comput. Appl. Math., 26 (1989), pp. 257–269.

[11] A. George and J. Liu, *Computer Solution of Large Sparse Positive Definite Systems*, Prentice-Hall, Englewood Cliffs, 1981.

[12] I. Gustafsson, *On first order factorization methods for the solution of problems with mixed boundary conditions and problems with discontinuous material coefficients*, Research Report 77.13R, Dept. of Computer Sciences, Chalmers Univ. of Technology and Univ. of Goeteborg, Goeteborg, Sweden, 1977.

[13] ———, *Modified incomplete Cholesky (MIC) methods*, in Preconditioning Methods. Theory and Applications, D. Evans, ed., Gordon and Breach, New York-London-Paris, 1983, pp. 265–293.

[14] F. Harary, *Graph Theory*, Addison-Wesley, Reading, 1969.

[15] Y. Notay, *Incomplete factorization of singular linear systems*, BIT, 29 (1989), pp. 682–702.

[16] ———, *Conditioning of Stieltjes matrices by S/P consistently ordered approximate factorizations*, BIT, (1990). submitted.

[17] ———, *Solving positive (semi)definite linear systems by preconditioned iterative methods*, these Proceedings, (1990).

[18] R. Varga, *On recurring theorems on diagonal dominance*, Linear Algebra Appl., 13 (1976), pp. 1–9.

[19] D. Young, *Iterative methods for solving partial differential equations of elliptic type*, Trans. Amer. Math. Soc., 76 (1954), pp. 91–111.

[20] ———, *Iterative solution of large linear systems*, Academic, New York, 1971.

On Some Parallel Preconditioned CG Schemes

R. Bramley H-C. Chen U. Meier A. Sameh *

September 2, 1990

1 Introduction

Over the last 40 years the theory of the CG algorithm has come to be well understood. However, the advent of modern computer architectures calls for a reexamination of and new approaches for CG methods. These approaches include careful utilization of the relationship of the algorithm to the machine architecture and the development of entirely new algorithms and preconditioners. In some case this leads to algorithms that are superior to previous sequential ones. This paper describes techniques developed for two shared memory multiprocessors. The first machine is the Alliant FX/8, which has 8 computational elements (CEs). Each CE has vector processing ability, with 8 vector registers containing 32 64-bit words each. Parallel execution of instructions is governed by a concurrency control bus, and the access to memory by the CE's is through a crossbar switch connected to a shared cache containing 16K 64-bit words. The maximum possible bandwidth between the cache and the CE's is 47 MW/sec. The cache in turn is connected to a 4MW main memory by a bus with a maximum possible bandwidth of 23.5 MW/sec. I/O and other noncomputational tasks are handled separately by 6 interactive processors which access memory through other caches. The second machine is Cedar, an experimental prototype machine being developed at the Center for Supercomputing Research and Development at the University of Illinois. Cedar has a global memory connecting up to 8 'clusters'. Each cluster is a modified Alliant FX/8, with a global interface connecting the Alliant processors to global memory through a two stage shuffle switch. The processors in a cluster still have access to cluster (local) memory through a shared cache, but other clusters cannot directly access that local memory.

The Alliant can use 2 levels of parallelism, if the problem is subdivided into parallel tasks each of which exploits the vector capabilities of the processors. Cedar can use three levels of parallelism; large grain by dividing the task among the clusters, medium grain by dividing the subtasks further among processors, and fine grain by using vectorization. Full utilization of these capabilities requires careful examination of the behaviour of the CG algorithm on the machines and innovative ways of applying it to new problems. This paper presents three such approaches. First, the performance of preconditioning methods is considered, and the tradeoffs in choosing methods on a

*Work supported by grants NSF-MIP-8410110, DOE DE-FG02-85ER25001, AT&T-AFFL-67-SAMEH, NSF-CCR-8717942, AFOSR-85-0211, and Digital Equipment Corp.

shared memory multiprocessor are examined. Secondly, an operator splitting technique for structural mechanics problems is described which creates parallel subtasks that can then be solved using the CG algorithm. This not only gives a preconditioner for CG that is suitable for multiprocessors, but also can give a superior sequential algorithm. Finally, a projection based scheme for solving nonsymmetric linear systems is presented which uses CG acceleration. This scheme provides a parallel implementation for multiprocessing, and leads to an algorithm that is more robust than other nonsymmetric solvers.

2 Preconditioning of CG

A suitable (left) preconditioner for CG is a matrix M such that the system $M^{-1/2}AM^{-1/2}(M^{1/2}x) = M^{-1/2}f$ requires fewer iterations to solve than $Ax = f$ does, and for which systems $Mz = r$ can be solved efficiently. The first property is independent of the machine used, while the second is highly machine dependent. The 'best' preconditioner is one which balances those two criteria in a way that minimizes the overall computation time.

For an example consider matrices A that arise when five point central differences are applied to self-adjoint second order elliptic PDE's with Dirichlet boundary conditions on an $n \times n$ grid. The first approach to consider is to use no preconditioner. In this case the basic operations required by CG are inner products ($\sigma = x^T y$), saxpy ($x \leftarrow x + \alpha y$), and the product of the five-diagonal matrix A and a vector. Clearly vectorization and parallelism can be utilized in all of these operations. However, for large problems the bandwidth to memory on the Alliant cannot supply data fast enough to stream data to and from all eight processors on any of these operations (see [1]). In this case the machine is not being used efficiently.

Using no preconditioning means that the second criterion above has been optimized at the expense of the first. At the other extreme, the block (M)ICCG preconditioners [2] [6] create matrices M which are the product of block triangular factors and can closely approximate A. Although they are effective in reducing the number of iterations required to solve the system, solving the resulting triangular systems is inherently recursive and not suitable for vectorization or parallelism. Some work has been done in approximating the inverses of the matrices of the linear systems to be solved and so replacing the highly recursive parts by vectorizable matrix-vector products at the cost of increasing the number of iterations (see, e.g., [17] and [13]).

One approach that can balance the two criteria is to use a polynomial $p(A)$ in A. In [14] it was shown that products $p(A)x$ can be formed with a partitioning scheme that reduces the number of data loads. The polynomial is obtained by minimizing some norm on the interval $[a, b]$, where a and b are estimates of the extremal eigenvalues of A obtained from a Gershgorin bound [16].

Another balancing approach is to use the point red-black ordering of the nodes to get

$$A = \begin{bmatrix} D_R & G \\ G^T & D_B \end{bmatrix},$$

where D_R and D_B are diagonal and G is a well structured sparse matrix with 5 non-zero

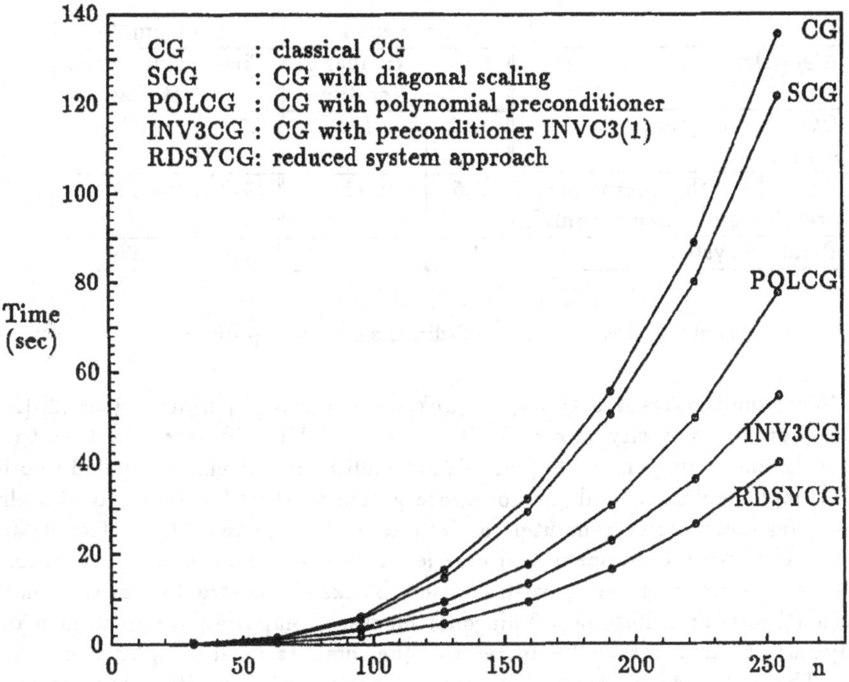

Figure 1: Times (in s) for solving the Poisson equation on an $n \times n$-grid

diagonals if n is even and 4 non-zero diagonals if n is odd. When scaled, the Schur complement of this system gives the reduced system $(I - H^T H)y = g$ of order $n^2/2$ for even n and of order $(n^2 - 1)/2$ for odd n where $H = D_R{}^{-1/2} G D_B{}^{-1/2}$. Once y is found, the solution x is trivial to retrieve from y. One advantage of this method is that now a system of half the order of the original one needs to be solved which often reduces the number of iterations. Also, the reduced matrix has nine non-zero diagonals in the case of odd n (eleven non-zero diagonals in the case of even n) and therefore a higher density. This improves data locality. For the results given in this paper, the reduced system was solved by CG with diagonal scaling, as the use of more complicated preconditioning schemes hardly decreases or even increases the time for solving the system on the considered machine (see [14]). Each of these and other approaches was thoroughly examined and tested on an Alliant FX/8 in [14]. The results show that in fact the performance of the considered schemes is severely limited by the inability of the machine to access data from memory fast enough to supply all of the processors. When ranked by the time required to solve a Poisson equation of order greater than 900, unpreconditioned CG requires the most time, followed in order by scaled, polynomial, INVC3(1) (see [13]), and reduced system preconditioning (see Figure 1). The savings possible is significant, with the reduced system approach requiring only about a third the time that unpreconditioned CG requires.

In the case of linear systems arising from the 7-point finite difference discretization of 3-dimensional self-adjoint elliptic P.D.E.'s, new approaches are possible that lead to larger vector lengths. An approach that has shown to be very efficient for the point

Algorithm	Problem 1 Time	Problem 1 Number of iterations	Problem 2 Time	Problem 2 Number of iterations
CG with diagonal scaling	12.2	176	21.9	192
ICCG(0) with hyperplane ordering and Eisenstat impl.	8.5	71	4.9	24
Reduced system	4.2	89	7.2	95

Table 1: Results for two 3-dimensional test problems

(M)ICCG preconditioners is ordering the unknowns across hyperplanes (see [2] [18]). For ten test problems of varying complexity, H. van der Vorst obtained the lowest timings for an Eisenstat implementation of this preconditioning scheme compared to other (M)ICCG implementations and the conjugate gradient method with diagonal scaling on various Japanese vector computers as well as on the Alliant FX/8. The reduced system approach for the 3-dimensional case leads to a matrix such as above where G is now a block tridiagonal matrix with diagonal blocks of the structure of G as in the 2-dimensional case and diagonal off-diagonal blocks. Computing the reduced matrix explicitly leads to an unreasonable increase of the computational complexity of the algorithm. The matrix-vector product is therefore performed implicitly which decreases the performance, even though we exploited the fact that parts of H can be kept in cache for the computation of $H^T Hx$. However, the timings for the considered test problems show that the reduced system approach is still about 2 to 3 times as fast as the conjugate gradient method with a diagonal scaling. Compared to the Eisenstat implementation of ICCG(0) with hyperplane ordering the reduced system is faster for 7 of the test cases on the Alliant FX/8. The results for the two problems in which the reduced system performs best or worst, respectively, compared to Van der Vorst's best ICCG scheme are given in Table 1. Problem 1 solves the self-adjoint problem with diffusion 1000 in the inner cubic region $(.1,.9) \times (.1,.9) \times (.1,.9)$ and 1 in the outer region of the unit cube for a mesh size of $60 \times 29 \times 13$. For problem 2, the diffusion is 0.1 on the cube with corners $(0,0,0)$ and $(.75,.75,.75)$, 100000 on the cube with corners $(.75,.75,.75)$ and $(1,1,1)$ and 1 elsewhere on the unit cube; the mesh size is $98 \times 28 \times 13$.

3 The SAS Method

A common class of problems in structural mechanics are isotropic 3D elasticity problems. When finite elements are used to discretize the differential equations governing the stresses, strains, and displacements of a 3D structure, a system of linear equations

$$Ax = f \tag{1}$$

results where A is usually the stiffness matrix, x the displacement vector, and f the force vector. The matrix A is often symmetric and positive definite and so the CG algorithm can be used to solve the system. However, as shown by [7, 8, 9], if symmetric elements are used in the finite element formulation along with a particular ordering of the nodes, the physical symmetry present in the problem can be exploited to decompose the problem into independent tasks. This approach, the symmetric and antisymmetric (SAS) method, relies on the existence of reflection matrices, i.e., symmetric signed permutation matrices. Let P be an $n \times n$ reflection matrix. Some important properties from [8, 9] are:

- Any vector $b \in \Re^n$ can be decomposed as $b = u + v$, with $u = Pu$ and $v = -Pv$. The vectors u and v are said to be symmetric and antisymmetric, respectively, with respect to P.
- Any matrix $A \in \Re^{n \times n}$ can be decomposed as $A = U + V$, with $U = PUP$ and $V = -PVP$. U and V are said to be reflexive and antireflexive, respectively, with respect to P.
- Given a nonsingular system $Ax = f$ with $A = PAP$, then x is symmetric (or antisymmetric) with respect to P iff f is symmetric (or antisymmetric) with respect to P.

Suppose that the structure, the boundary conditions, and the material properties all are symmetric with respect to at least one coordinate axis. In [9] it is shown that A in (1), via proper discretization and the SAS ordering of unknowns, is reflexive with respect to a reflection matrix

$$P = \begin{bmatrix} 0 & G \\ G & 0 \end{bmatrix} \tag{2}$$

where G is a signed identity matrix of order k, with $n = 2k$. Define

$$X = \frac{1}{\sqrt{2}} \begin{bmatrix} I & -G \\ G & I \end{bmatrix}, \quad \tilde{A} = X^T A X, \quad \tilde{x} = X^T x, \quad \tilde{f} = X^T f. \tag{3}$$

Since A is reflexive with respect to P, it is easily shown that

$$\tilde{A} = \begin{bmatrix} A_{11} + A_{12}G & 0 \\ 0 & A_{22} - A_{21}G \end{bmatrix}, \quad \text{where} \quad A = \begin{bmatrix} A_{11} & A_{12} \\ A_{21} & A_{22} \end{bmatrix}. \tag{4}$$

The problem is thus decomposed into two completely independent subsystems (see also [7]). If the problem is physically symmetric about two coordinate axes, then $A_{11} + A_{12}G$ and $A_{22} - A_{21}G$ are also reflexive, with respect to another reflection matrix H. The decomposition can be carried out further to yield four independent subsystems. This procedure can be continued recursively until no disjoint submatrices possess the SAS property.

Equally important, the similarity transformation matrix X is orthogonal, and the submatrices generated are symmetric and positive definite. Hence the CG algorithm can be used to solve those independent subsystems. Viewed in this manner, the SAS decomposition can be seen as a preconditioner for CG. However, symmetry of the structure, boundary conditions, and discretization is required even though the stiffness matrix is symmetric and positive definite when those conditions do not hold. An explicit example demonstrating this is given below.

$T_{1,1}$	$T_{1,2}$	$T_{1,4}$	$T_{1,8}$	$T_{8,8}$
154.1	76.2	32.3	18.5	4.5

Table 2: CG Times for Bar B1

First, consider a 3D isotropic prismatic long bar B_1 of dimensions $(L, b, c) = (30.0, 1.5, 2.0)$ with Young's modulus $E = 1000$ and Poisson's ratio $\nu = 0.3$. The center of the bar is placed at $(15, 0, 0)$ and the bar is fixed at both ends, that is, the displacements in all three directions on the planes $x = 0$ and $x = L$ are zero. A force $q = 200$ is applied vertically 2/3 of the way down the length of the bar. Since the problem is physically symmetric about all three coordinate axes, it can be decomposed into 8 subproblems by the SAS method when a basic 8-node rectangular hexahedral finite element is used with the SAS ordering of the nodes. A $15 \times 7 \times 7$ mesh is imposed on the bar. The decomposition is done using the planes $y = 0$, $z = 0$, and $x = 15$ in order. Within each subdomain the natural ordering of nodes, plane by plane, is used starting with the $x = 30$ plane. An important computational feature is that the assembled system stiffness matrix is not decomposed, but each element matrix is decomposed and then assembled into the corresponding subsystems.

The CPU time in seconds on an Alliant FX/8 required by the CG algorithm is measured. Because of its effectiveness on multiprocessors as shown in section 2, main diagonal preconditioning is applied. The problem is decomposed into $N_s = 1, 2, 4,$ or 8 subproblems using the SAS method (1 subproblem means that (1) is solved directly), and parallel computations are performed within each linear subsystem. The starting estimate is $x = 0$ and the stopping test is $\| r_k \| \leq 10^{-6} \| f \| / \sqrt{N_s}$ where r_k is the residual on the kth iteration. Table 2 shows the results for execution on $m = 1, 2, 4, 8$ processors where $T_{i,j}$ stands for the CPU time using i processors when the problem is decomposed into j subproblems. The CPU time is reduced by a factor of more than 32 by going from one subdomain using one processor to eight subdomains using all eight processors of the Alliant. This results from a combination of the SAS decomposition and parallelism. The speed improvements due to the SAS decomposition are given by the first four columns of Table 2. These improvements are achieved mainly from the reduction of the serial computational work. For this problem, the SAS decomposed subproblems have the same number of nonzero diagonals, even though the number of subproblems is increasing. Speedups resulting from the use of parallelism afforded by the Alliant FX/8 can be seen from the last two columns of the table. Readers are referred to [7, 9] for more comparisons.

For physically nonsymmetric problems the SAS method can be used to create a preconditioner for the stiffness system by using the property that any matrix can be decomposed into a reflexive and antireflexive part $A = U + V$ or into a reflexive part U plus its remainder W, i.e., $A = U + W$. If U is dominant in some sense, it can be used as a preconditioner for A. As an example consider bar B2, which has the same dimensions and material properties as B1, but is fixed at the left end and supported by two linear springs at locations $(L, \pm b/2, c/2)$ on the free end. The spring elastic constants are taken as $K_1 = 100$ and $K_2 = 50$, respectively. A uniform simple bending moment

$M = 60$ is applied across the cross section at the right end, and a concentrated force $P = 10$ is applied at $(L, -b/2, c/2)$. The presence of the two springs prevents the global stiffness matrix from being SAS decomposed. Without the springs, the problem would be decomposable into 4 subproblems. Based on this observation, we split the matrix $A = U + W$, where W contains the stiffness contribution from the springs and U consists of the rest. Then U can be used as a preconditioner when the CG algorithm is used to solve the system (1). The resulting algorithm is called PCGSAS, and is compared to CG with no preconditioning (CGNOPC), CG with diagonal preconditioning (PCGDIA), and a direct banded solver (GEDROW). The preconditioner U for PCGSAS is decomposed once at the start of the algorithm during the assembly of the stiffness matrix A, and a direct banded solver is used to solve the systems $Ud_k = r_k$ needed on each iteration of the preconditioned CG algorithm. When the timings are compared, PCGSAS is about 26, 19, and 6.5 times as fast as CGNOPC, PCGDIA, and GEDROW, respectively, when 8 processors are used (see [7] for more information).

As the first example shows, developing an implementation of the CG algorithm for a shared memory multiprocessor not only can lead to a parallel algorithm, but also one that is superior to the usual sequential algorithm. As the second example shows, the effort also can lead to a highly effective preconditioner.

4 Block Kaczmarz Schemes

One of the outstanding problems in linear system is finding solvers that can handle large sparse nonsymmetric systems $Ax = f$ with indefinite symmetric parts and eigenvalues possibly lying in all four quadrants of the complex plane. These conditions can cause failure in Chebyshev methods and Krylov subspace methods such as Gmres(k). Although preconditioning sometimes can convert such problems into more tractable ones by, e.g., making the symmetric part positive definite, there is no assurance that this will happen or even that the preconditioner will exist. CG applied to the normal equations (CGNE) theoretically converges for any nonsingular system, but implicitly squares the condition number of the system. In practice this can cause failure, and again standard preconditioners such as ILU may not exist and may not help even when they do exist.

A solver which shares CGNE's theoretical robustness but has superior numerical robustness is the CG accelerated block Kaczmarz scheme. Partition the rows of A into m blocks $A_1^T, A_2^T, \ldots, A_m^T$ and the rows of f conformally. The ith block of equations $A_i^T x = f_i$ defines a linear manifold, and the solution x^* is the intersection point of those m manifolds. Applying successive projections gives the iteration

$$x_{k+1} = (I - P_m) \cdots (I - P_2)(I - P_1)x_k + \hat{f}, \tag{5}$$

where $\hat{f}$ is a modified right hand side vector and $P_i = A_i(A_i^T A_i)^{-1} A_i^T$ is the orthogonal projector onto the range of A_i. This approach was first proposed by S. Kaczmarz [10] for the case where each block row is a single row of A. Because of the slow convergence of (5), A. Björck and T. Elfving [3] proposed following a forward sweep through the block rows by a backward sweep and using CG acceleration on the resulting system $(I - Q)x = \bar{f}$, where $\bar{f}$ is another modified right hand side vector and

$$Q = (I - P_1)(I - P_2) \cdots (I - P_m) \cdots (I - P_2)(I - P_1).$$

Since matrix-vector products with Q require solving $2m-1$ linear least squares problems, the key to an efficient implementation is parallelism. As an example, suppose A is obtained from seven point central differencing of a non-selfadjoint PDE on the unit cube with Dirichlet boundary conditions and an $n \times n \times n$ mesh imposed. Nodes separated by at least two other nodes in the mesh have associated equations that share no unknowns. Similarly, lines of nodes in the mesh separated by at least two other lines of nodes have associated blocks of equations that have no unknowns shared between the blocks. If such disjoint blocks are grouped together to form one of the A_i^T's, the corresponding projection P_i can be computed as independent subtasks. Each subtask consists of computing the projection onto the range of a set of rows of A corresponding to a single line of nodes in the mesh.

A simple way to do this is to group every third line on every third plane into the same A_i^T. This gives 9 block rows, each consisting of $n^2/9$ subblocks that can be treated in parallel when computing the projections. Each subblock has the form

$$C^T = [D_1, 0, \ldots, 0, D_2, T, D_3, 0, \ldots, 0, D_4]$$

with D_i diagonal and T tridiagonal matrices of order n. The $n^2/9$ subtasks take the form

$$y = x - CL^{-T}DL^{-1}C^T x \tag{6}$$

with L unit lower triangular and D diagonal such that $LD^{-1}L^T = C^T C$. A similar scheme was proposed for 2D in [12] and 3D in [4]. Both demonstrated the numerical robustness of the CG accelerated block-Kaczmarz method by comparisons with GMRES(k) with and without (M)ILU preconditioning and with CGNE with and without ILU preconditioning. As an example, consider the PDE

$$\Delta u - 1000x^2 u_x + 1000u = F \tag{7}$$

with Dirichlet boundary conditions chosen so that the exact solution is

$$u(x, y, z) = e^{xyz} \sin(\pi x) \sin(\pi y) \sin(\pi z)$$

Table 3 shows the results for block-Kaczmarz, GMRES(10), and CGNE with a stopping test of $\| Ax_k - f \|_2^2 \leq 10^{-9}$. GMRES(10) with and without (M)ILU preconditioning stalls during the iteration by not reducing the residual norm by a difference of 10^{-6} over the course of 10 iterations, and in this case the iteration is stopped. When allowed to continue, it runs to the maximum allowed number of iterations without further reducing either the residual or error norm. Furthermore, when allowed to run for the same amount of time as block Kaczmarz all of the other methods still fail to achieve a comparable residual size. Experiments with GMRES(k) with $k = 20$, 30, and 40 fail in the same way. CGNE with and without ILU(0) preconditioning reaches the maximum allowed number of iterations without sufficiently reducing the residual norm, and continues to fail to do so even after 14000 iterations. Table 4 shows the failure rates over 9 test problems on 3 mesh sizes. On the one case where block-Kaczmarz fails, so did every other solver tested.

Furthermore, the parallel tasks (6) can be implemented using vectorization when multiplying by C, D, and C^T. Even though the solves with L and L^T are inherently

	Block Kaczmarz	GMRES No Prec	GMRES ILU	GMRES MILU	CGNE No Prec	CGNE ILU
Iter	2656	175	98	142	8001	4001
Time	369.8	20.2	62.0	89.1	981.7	2337.0
Resid	1.5E-5	6.1E+0	3.7E+0	1.8E+2	1.5E-3	8.4E-2
Error	1.2E-2	2.5E+0	8.4E+1	5.2E+1	2.0E+0	2.5E+1
Failure		*	*	*	*	*

Table 3: Performance of Methods for Sample Problem

Block Kaczmarz	GMRES(10) No Prec	GMRES(10) ILU	GMRES(10) MILU	CGNE No Prec	CGNE ILU
3.7%	33.3%	51.9%	74.1%	25.9%	59.3%

Table 4: Failure Rates for Six Linear Solvers

sequential, $n^2/9$ of them are performed in parallel, unlike the triangular solves necessary for (M)ILU which are performed sequentially. For this reason block-Kaczmarz is comparable with or even faster than the other methods on the Alliant FX/8. For a machine like Cedar with three levels of parallelism, a related row projection method which requires the sum instead of the product of projections on each iteration can be used; then each cluster can be assigned a projection, which can be computed with parallelism and vectorization as described above (see [4] for details about this approach).

Block-Kaczmarz can be viewed as a preconditioner for the CG method, one which like CGNE transforms a nonsymmetric problem into a symmetric one. Unlike CGNE, block-Kaczmarz can actually improve the spectral properties of the matrix. Although initially examined because of its suitability for multiprocessors, the resulting algorithm can be more robust than commonly used sequential algorithms.

5 Summary

Efficient use of CG on shared memory multiprocessors requires the detailed examination and testing of the algorithm and its preconditioners as well as the development of innovative approaches to preconditioning. In some cases these approaches lead to improved algorithms even for sequential machines.

References

[1] W. Abu-Sufah and A. Malony, Experimental results for vector processing on the Alliant FX/8, *CSRD Tech. Rept. 539*, Center for Sueprcomputing Research and Development, Univ. Illinois-Urbana, (1986).

[2] O. Axelsson and V. Eijkhout, Robust vectorizable preconditioners for three-dimensional elliptic difference equations with anisotrophy, *Algorithms and Applications on Vector and Parallel Computers* (Ed. H.J.J. te Riele, Th.J. Dekker and H.A. van der Vorst), Elsevier Science Publishers, 1987, pp.279-306.

[3] A. Björck, T. Elfving, Accelerated projection methods for computing pseudo-inverse solutions of systems of linear equations, *BIT*, 145-163, 19(1979).

[4] R. Bramley, Row projection methods for linear systems, *CSRD Tech. Rept. 881*, Center for Supercomputing Research and Development, Univ. Illinois-Urbana, (1989).

[5] R. Bramley, A. Sameh, A robust parallel solver for block tridiagonal systems, *CSRD Tech. Rept. 806*, Center for Supercomputing Research and Development, Univ. Illinois-Urbana, (1988).

[6] P. Concus, G. Golub, and G. Meurant, Block preconditioning for the conjugate gradient method, *SIAM J. Sci. Stat. Comput.*, vol. 6, pp. 220-252, (1985).

[7] Hsin-Chu Chen and Ahmed Sameh, A domain decomposition method for 3D elasticity problems, *Applications of Supercomputers in Engineering: Fluid Flow and Stress Analysis Applications* (Ed. C.A. Brebbia and A. Peters), Computational Mechanics Publications, Southampton University, UK, Sep. 1989, pp.171-188.

[8] Hsin-Chu Chen and Ahmed Sameh, A matrix decomposition method for orthotropic elasticity problems, *SIAM Journal on Matrix Analysis and Applications*, vol. 10, No. 1, pp.39-64, Jan. 1989.

[9] Hsin-Chu Chen, The SAS domain decomposition method for structural analysis, *CSRD Tech. Rept. 754*, Center for Supercomputing Research and Development, Univ. Illinois - Urbana, (1988).

[10] S. Kaczmarz, Angenäherte Auflösung von Systemen linearer Gleichungen, *Bull. intern. Acad. polonaise Sci. lettres (Cracouie); Class sci. math. natur.: Seira A. Sci. Math.*, 355-357 (1939).

[11] C. Kamath, Solution of nonsymmetric systems of equations on a multiprocessor, *CSRD Tech. Rept. 591*, Center for Supercomputing Research and Development, Univ. Illinois - Urbana, (1986).

[12] C. Kamath, A. Sameh, A projection method for solving nonsymmetric linear systems on multiprocessors, *Parallel Computing*, 291-312, vol. 9 (1988/1989).

[13] G. Meurant, The block preconditioned conjugate gradient algorithm on vector computers, BIT, vol. 24, pp. 623-633, (1984).

[14] U. Meier and A. Sameh, The behavior of conjugate gradient algorithms on a multivector processor with a hierarchical memory, *J. Comp. App. Math.*, pp. 13-32, vol. 24, (1988).

[15] J. Meijerink and H. van der Vorst, An iterative solution method for linear systems of which the coefficient matrix is a symmetrix M-matrix, *Math. Comput.* 31(137), pp. 148-162, 1977.

[16] Y. Saad, Practical use of polynomial preconditionings for the conjugate gradient method, *SIAM J. Sci. Stat. Comput*, vol. 6, pp. 865-881, (1985).

[17] H. van der Vorst, The performance of Fortran implementations for preconditioned conjugate gradients on vector computers, *Parallel Computing*, vol. 3, pp. 49-58, (1986).

[18] H. van der Vorst, ICCG and related methods for 3D problems on vectorcomputers, *Report No. A-18*, Data Processing Center, Kyoto University, Kyoto, Japan (1987).

Preconditioning indefinite systems arising from mixed finite element discretization of second-order elliptic problems

Richard E. Ewing, Raytcho D. Lazarov*, Peng Lu,
Panayot S. Vassilevski*

Department of Mathematics, Box 3036, University Station
University of Wyoming, Laramie, Wyoming 82071, USA

Abstract

We discuss certain preconditioning techniques for solving indefinite linear systems of equations arising from mixed finite element discretizations of elliptic equations of second order. The techniques are based on various approximations of the mass matrix, say, by simply lumping it to be diagonal or by constructing a diagonal matrix assembled of properly scaled lumped element mass matrices. We outline two possible alternatives for preconditioning. One can precondition the original (indefinite) system by some indefinite matrix and hence use either a stationary iterative method or a generalized conjugate gradient type method. Alternatively as in the particular case of rectangular Raviart-Thomas elements, which we consider, one can perform iterations in a subspace, eliminating the velocity unknowns and then considering the corresponding reduced system which is elliptic. So we can use the ordinary preconditioned conjugate gradient method and any known preconditioner (of optimal order, for example, like the multigrid method) for the corresponding finite element discretization of the elliptic problem. Numerical experiments for some of the proposed iterative methods are presented.

Keywords: indefinite system, preconditioning, iterations in subspace, conjugate gradients, mixed finite elements, second order elliptic problems.

Subject Classifications: AMS(MOS) 65F10, 65N20, 65N30.

---

* On leave from Center for Informatics and Computer Technology, Bulgarian Academy of Sciences, G. Bontchev str., bl. 25-A, 1113 Sofia, Bulgaria

1. Introduction

In this paper, we consider indefinite linear systems of equations that arise in the mixed finite element discretization of second-order elliptic problems. In general, this approximation leads to systems of the form

$$AX = \begin{pmatrix} M & N \\ N^T & 0 \end{pmatrix} \begin{pmatrix} U \\ P \end{pmatrix} = \begin{pmatrix} 0 \\ -F \end{pmatrix}, \tag{1}$$

where M is a symmetric, positive definite, and constantly-conditioned matrix and A is nonsingular.

In the existing literature, one can find various approaches for solving such systems. For example, in [6], Bramble and Pasciak have constructed a preconditioner for a properly transformed form of the original system where the resulting matrix is definite, and the preconditioned conjugate gradient method is then applied. Another approach was proposed and studied in Bank, Welfert and Yserentant [5], where a parameter-free iterative method (but not of conjugate gradient type) for solving the original system is used. However, at every iteration, one uses a variable-step preconditioner, which involves approximate solving of systems with M and approximate solving (possibly with conjugate gradient-method) of the approximate reduced system

$$N^T \widetilde{M}^{-1} N\, P = F, \tag{2}$$

where $\widetilde{M}$ is an approximation to M.

A general approach of parameter-free preconditioned conjugate gradient (PCG) type methods for solving problem (1) was proposed recently in Axelsson and Vassilevski [4].

For certain applications, the cost of evaluating M^{-1} applied to a vector is not too expensive. This is true for tensor product Raviart-Thomas elements on a regular rectangular grid, where the action of M^{-1} involves banded solution (of bandwidth proportional to the degree r of the piece-wise polynomials used in the finite element spaces). For such situations one can consider the (exactly) reduced system

$$N^T M^{-1} N\, P = F, \tag{3}$$

which has a symmetric and positive definite matrix.

In this case in Bramble, Pasciak and Xu [7], an optimal order V-cycle multigrid method has recently been proposed. In the present paper we choose a preconditioner B for the original matrix A of the following form:

$$B = \begin{pmatrix} \widetilde{M} & 0 \\ N^T & -\widetilde{C} \end{pmatrix} \begin{pmatrix} I & \widetilde{M}^{-1}N \\ 0 & I \end{pmatrix} = \begin{pmatrix} \widetilde{M} & N \\ N^T & \widetilde{D} \end{pmatrix}, \tag{4}$$

where $\widetilde{D} = N^T \widetilde{M}^{-1} N - \widetilde{C}$, and the matrices $\widetilde{M}$ and $\widetilde{C}$ (or $\widetilde{D}$) have to be specified.

We concentrate mainly on the case $\widetilde{D} = 0$, that is exact solution of the perturbed reduced system (2). Approximate solving of (2) (that is $\widetilde{C} \neq N^T \widetilde{M}^{-1} N$, or equivalently $\widetilde{D} \neq 0$) can be considered in the framework of the papers of Bank, Welfert and Yserentant [5], Axelsson and Vassilevski [4] or Ewing and Wheeler [8].

A simple choice for $\widetilde{M}$, $\widetilde{M} = \omega I$, was made in Lu, Allend and Ewing [11], and the parameter was chosen globally in order to minimize the norm of the iteration matrix $I - B^{-1}A$ in a stationary iterative method. Then, the system with the matrix $\widetilde{C} = \omega^{-1} N^T N$ was solved using a multigrid method for the Poisson equation. As shown in [11], for rapidly varying coefficients of the differential equation, this iterative method may converge very slowly although being independent of the mesh parameters. In this paper, we advocate another diagonal approximation $\widetilde{M}$ to M, which is obtained by assembling properly scaled lumped element mass matrices. Then the resulting reduced system (2) is solved by inner iterations. We give a precise upper bound of 1/2 for the spectral radius of the iteration matrix $I - B^{-1}A$. We also prove that the eigenvalues of $B^{-1}A$ are real and positive and that they are dominated by the eigenvalues of $\widetilde{M}^{-1}M$. Hence the generalized conjugate gradient method from Axelsson [2] will have a convergence factor bounded by $(\sqrt{\varkappa}-1)/(\sqrt{\varkappa}+1)$, where $\varkappa$ is the condition number of $B^{-1}A$. We consider this to be the main result of this paper.

The remainder of the paper is organized as follows. In Section 2, we formulate the mixed finite element discretization of second-order elliptic problems using Raviart-Thomas elements. In the next section, we give some properties of the finite element matrix. In Section 4, the preconditioning technique sketched above is described and the corresponding main results are proved. At the end, we present some numerical experiments using the proposed preconditioner in a stationary iterative procedure.

2. Problem Formulation

Consider the following equation governing steady state heat transfer in a nonhomogeneous d-dimensional conducting medium:

$$\sum_{i,j=1}^{d} \frac{\partial}{\partial x_i}\left(k_{ij}\frac{\partial p}{\partial x_j}\right) = f \text{ in } \Omega. \tag{5}$$

We suppose that the matrix $K = \{k_{ij}(x)\}_{i,j=1,\dots,d}$ is symmetric and uniformly positive definite in $\overline{\Omega}$; i.e., there exist constants $\alpha_1, \alpha_0 > 0$, such that

$$\alpha_0 \xi^T \xi \leq \xi^T K \xi \leq \alpha_1 \xi^T \xi, \qquad \forall \xi \in \mathbf{R}^d. \tag{6}$$

Therefore, introducing the vector

$$\mathrm{u} = -K\nabla p, \text{ where } \nabla p = \left(\frac{\partial p}{\partial x_1}, \dots, \frac{\partial p}{\partial x_d}\right)^T, \tag{7}$$

equation (5) reads as

$$-\text{div } \mathbf{u} = f \text{ in } \Omega. \tag{8}$$

For simplicity, the domain Ω is taken to be the unit hypercube in $\mathbb{R}^d$; i.e. $\Omega = [0,1]^d$. We impose the boundary condition $p = 0$ on $\partial\Omega$.

In the context of fluid of a single phase in a d-dimensional porous medium, $\mathbf{u}, p$, and f represent the Darcy velocity, pressure, and source term, respectively. Problem (7) and (8) can be formulated as a saddle point problem. Let us first introduce the vector space

$$\mathbf{V} = \{\mathbf{v} = (v_1, \ldots, v_d) : v_i \in L^2(\Omega),\ \text{div } \mathbf{v} \in L^2(\Omega)\}$$

and endow it with the norm

$$\|\mathbf{v}\|_{\mathbf{V}}^2 = \|\mathbf{v}\|_{H(\text{div};\Omega)}^2 = \|\mathbf{v}\|_{L^2(\Omega)}^2 + \|\text{div } \mathbf{v}\|_{L^2(\Omega)}^2.$$

For $\mathbf{u}, \mathbf{v} \in \mathbf{V}$ and $p, w \in L^2$, we define the bilinear forms

$$\begin{aligned} &\text{(a)} \quad a(\mathbf{u}, \mathbf{v}) = (K^{-1}\mathbf{u}, \mathbf{v}), \\ &\text{(b)} \quad b(\mathbf{u}, w) = (\text{div } \mathbf{u}, w). \end{aligned} \tag{9}$$

Then problem (7) and (8) is equivalent to solving the saddle point problem given by

$$\begin{aligned} a(\mathbf{u}, \mathbf{v}) - b(\mathbf{v}, p) &= 0, && \forall \mathbf{v} \in \mathbf{V}, \\ -b(\mathbf{u}, w) &= -(f, w), && \forall w \in L^2. \end{aligned} \tag{10}$$

Let T_h be a partitioning of the domain Ω by parallelipiped finite elements and $\mathbf{V}_h^r$ and W_h^r be the Raviart-Thomas finite element spaces, which are subspaces of $\mathbf{V}$ and L^2, respectively. The spaces $\mathbf{V}_h^r$ and W_h^r are usually referred to as Raviart-Thomas spaces of order r, [12], for $r \geq 0$.

A Lagrangian basis for these spaces that is useful for implementation is constructed using a tensor product of Gaussian and Lobatto points. An example of sets of such points in the reference element $E = [-1,1]^d$ is shown in Figure 1 ($r = 0$, $d = 2$ and $r = 1$, $d = 2$).

Now, we formulate the mixed finite element approximation to problem (10). Find $\mathbf{U} \in \mathbf{V}_h^r$ and $P \in W_h^r$ satisfying:

$$\begin{aligned} a(\mathbf{U}, \mathbf{V}) - b(\mathbf{V}, P) &= 0, && \mathbf{V} \in \mathbf{V}_h^r \\ -b(\mathbf{U}, w) &= -(f, w), && w \in W_h^r. \end{aligned} \tag{11}$$

Since The Raviart-Thomas spaces satisfy the Babuska-Brezzi condition [9,12], the finite element solution $(\mathbf{U}, P)$ exists and converges in $V \times L^2$-norm to the exact solution with a rate $O(h^{r+1})$, [9].

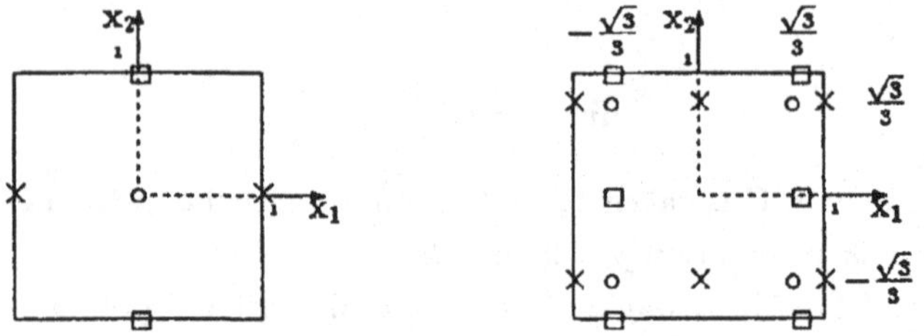

$r = 0$ $\qquad\qquad\qquad$ $r = 1$

× – U_1, $\quad$ □ – U_2, $\quad$ ∘ – P

Figure 1

Raviart-Thomas elements, $r = 0$ and $r = 1$

3. Properties of the Finite Element Matrix

We concentrate our attention on methods for solving the corresponding system of linear equations for the unknown values of U and P at the grid points. We introduce the same notation for these vectors: U, vector of the values of U and P, the values of P. Then, (11) can be presented in the algebraic system

$$\begin{aligned} M\mathbf{U} + NP &= 0, \\ N^T\mathbf{U} &= -F, \end{aligned} \tag{12}$$

where the matrices M and N are defined by the bilinear forms $a(\cdot,\cdot)$ and $b(\cdot,\cdot)$, respectively.

We now consider the case $d = 2$. If we partition the unknown values of U as $\mathbf{U} = (U_1^T, U_2^T)^T$, then the matrix M will have the following 2×2 block form

$$M = \begin{pmatrix} M_{11} & M_{12} \\ M_{12} & M_{22} \end{pmatrix}. \tag{13}$$

In the particular case when $k_{12}(x) \equiv k_{21}(x) \equiv 0$, we have

$$M = \begin{pmatrix} M_{11} & 0 \\ 0 & M_{22} \end{pmatrix}. \tag{14}$$

Then M_{ii}, $i = 1, 2$ will have a block diagonal structure, and one can easily eliminate the velocity vector U from system (12). We obtain

$$N^T M^{-1} N\, P = F. \tag{15}$$

One can now see that the matrix of the reduced system (15) is symmetric and positive definite. In general, this is an approximation of Equation (5) that has quite a strange global stencil (see Figure 2).

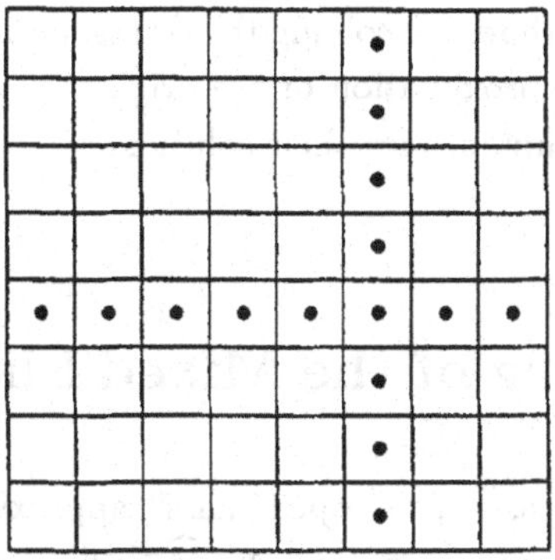

Figure 2
Stencil for the pressure equation, $r = 0$.

The matrix M is obtained using exact computation (never done in practice for variable coefficients) or quadrature rules that are exact for polynomials of degree $2r+3$ in each variable. If we use Lobatto quadrature in the x_i direction with $r+2$ nodes and Gauss quadrature in the remaining directions for computing M_{ii}, then we obtain a diagonal matrix $\widehat{M}_{ii}$; i.e.,

$$(k_{ii}^{-1}U_1, V_1) \cong V_1^T \widehat{M}_{ii} U_1. \tag{16}$$

In this case, the system (15) is an immediate consequence of our computational procedure. In the case $r = 0$, this is a 5-point approximation for the pressure equation.

Therefore, we have the following three cases:

(i) the general system (12) with a "mass" matrix M in the form (13);

(ii) the system (12) with a consistent "mass" matrix M in the form (14) where M_{ii}, $i = 1, 2$, have block diagonal form with banded blocks and are therefore easily (rather effectively) reducible to the system (15);

and

(iii) the "lumped mass' system, written in the form

$$\begin{aligned} \widehat{M}\mathbf{U} + NP &= 0, \\ N^T\mathbf{U} &= -F, \end{aligned} \tag{17}$$

which yields $N^T \widehat{M}^{-1} N\, P = F$.

It is important to mention that: (a) in general, the system of case (i) cannot be reduced to the form (15) in an effective way; (b) the reduction of system (12) with a matrix M of the form (14) can be performed in an effective way (the arithmetric operations are proportional to the number of unknowns but the matrix $N^T M^{-1} N$ is given only in an implicit way); and (c) the "lumped mass" system of (iii) is given in an explicit way and can be obtained using an element by element assembling procedure.

Further, we discuss methods for solving the corresponding system (12), arising from the mixed finite element approximation of the original problem (10). We now discuss various preconditioning techniques for their solution.

4. Preconditioning of the Mixed Finite Element Systems

We first consider the case of "lumped mass" approximation: the system can be written in the form (17), where the matrix $N^T \widehat{M}^{-1} N$ is known explicitly. For the lowest order finite elements, $r = 0$, this is a 5-point approximation of the elliptic equation, and here we can use some of the known fast solvers. For example, a multigrid method will be suitable even when the coefficients are piecwise constant or piecewise continuous. Therefore, we can assume that problem (17) can be solved in an efficient way. For the elements of higher order ($r > 0$), there is no obvious fast solver for the reduced system (17). Probably, one can design a multigrid method for this system or use some preconditioning method with preconditioners based on the approximations of lower order. This approach needs further investigation and computational experiments. Anyway, we may suppose that we have constructed a fast solver for the problem (17), for any r, in a rectangular domain. Further, we use this solver to design iterative methods for the general problem (12).

In the general preconditioning matrix B, we can distinguish two cases: (a) $\widetilde{M} = M$ and (b) $\widetilde{M} \neq M$. The first case, $\widetilde{M} = M$, is easier to analyze since all iterations are performed in the subspace of vectors satisfying the condition $MU + NP = 0$ and where the matrix A is definite. In many cases, the realization of this approach can be much more difficult since we need to evaluate the action of M^{-1}. In our case of tensor product elements on a refular rectangular grid, this evaluation involves banded solvers (of bandwidth proportional to r) along the lines $0 < x_i < 1$ for the velocity component U_i and therefore can be done with arithmetic work proportional to the number of unknowns. In the case of triangular elements, one might consider iterative evaluation of M^{-1}, which is a well-conditioned matrix. Since A is a symmetric matrix in this subspace, we can use the preconditioned conjugate gradient method.

In order to better explain the preconditioner in this case, we present matrix A in a factored form:

$$A = \begin{pmatrix} M & N \\ N^T & 0 \end{pmatrix} = \begin{pmatrix} M & 0 \\ N^T & -N^T M^{-1} N \end{pmatrix} \begin{pmatrix} I & M^{-1} N \\ 0 & I \end{pmatrix}.$$

Here $C = -N^T M^{-1} N$ is the Schur complement of A after elimination of the velocity vector U. Since solving the system with matrix C is a difficult problem (for arbitrary r), we could solve it by preconditioning. This will produce a preconditioner B, where $\widetilde{M} = M$ in (4).

In this case, we have a variety of possibilities of approximating C by $\widetilde{C}$, or equivalently approximating M^{-1}. For example, we may choose $\widetilde{C} = N^T \widehat{M}^{-1} N$, where $\widehat{M}$ is the lumped mass matrix M or $\widehat{M}$ is the diagonal part of M. These two choices are particular cases of the general approach of replacing M by some variant of its incomplete factorization [1]. In all cases considered, matrix $\widetilde{C}$ is spectrally equivalent to A (in the subspace $M\mathbf{U} + NP = 0$) with constants independent of the problem size. Thus, this preconditioner B is an optimal one. Nevertheless, in this case we still relay on the existence of an effective solver of the problem $\widetilde{C}P = F$ where the condition number of $\widetilde{C}$ is $O(h^{-2})$. This can be done by a multigrid method [7] or an algebraic multilevel method [3], for instance.

The second case, $\widetilde{M} \neq M$, is much more general, since we are predonditioning the original matrix A and need not exact solving with M. It can be applied for general differential equations and mixed approximations of type (12). Here we must overcome two main difficulties: (i) to choose a matrix $\widetilde{M}$ such that it scales M in a proper way and (ii) to perform the convergence analysis of the iterative procedure; in this case, the spectrum of $B^{-1}A$ is complex in general (if $\widetilde{D} \neq 0$) and difficult to analyze.

This approach was used by Lu, Allen and Ewing [11] with $\widetilde{M} = \omega I$ and $\widetilde{D} = 0$, where the choice of the iteration parameter ω is based on the properties of M. The computations performed in [11] showed that the iterative method slows down substantially when $\max_{\Omega} k_{ii}(x) / \min_{\Omega} k_{ii}(x) \gg 1$. This means that I does not scale the matrix M in a proper way.

A better choice for $\widetilde{M}$ is a diagonal matrix obtained from assembling properly scaled lumped element mass matrices, which we describe later. In the analysis below we treat the equation (5) without mixed derivatives, that is $M_{12} = M_{21} = 0$, and $k_{11}(x) = k_{22}(x) = k(x)$, and consider only the case $\widetilde{D} = 0$.

We have the following main result:

Theorem 1. The eigenvalues of $B^{-1}A$ are real and positive and lie in the interval $[\lambda_{\min}, \lambda_{\max}]$ where $\lambda_{\min}, \lambda_{\max}$ are the extreme eigenvalues of $\widetilde{M}^{-1}M$.

Proof. Consider the following generalized eigenvalue problem

$$A\begin{pmatrix}\mathbf{U}\\ P\end{pmatrix} = \lambda B\begin{pmatrix}\mathbf{U}\\ P\end{pmatrix}, \qquad \text{(see (4) with } \widetilde{D} = 0\text{)},$$

or equivalently,

$$\begin{aligned}(M - \lambda\widetilde{M})\mathbf{U} + NP &= \lambda NP,\\ N^T\mathbf{U} &= \lambda N^T\mathbf{U}.\end{aligned} \tag{18}$$

Hence either $\lambda = 1$ or $N^T\mathrm{U} = 0$. If $\lambda = 1$, then $(\widetilde{M} - M)\mathrm{U} = 0$; that is $\lambda = 1$ is also an eigenvalue of $\widetilde{M}^{-1}M$. Now let $N^T\mathrm{U} = 0$. Then multiplying the first equation in (18) by U^{*T} (the complex conjugate transpose of U) we get

$$\mathrm{U}^{*T}(M - \lambda\widetilde{M})\mathrm{U} + (N^T\mathrm{U})^{*T}P = \lambda(N^T\mathrm{U})^{*T}P.$$

That is $\mathrm{U}^{*T}(M - \lambda\widetilde{M})\mathrm{U} = 0$ since $N^T\mathrm{U} = 0$. Hence λ is real and positive and since

$$\lambda = \frac{\mathrm{U}^{*T}MU}{\mathrm{U}^{*T}\widetilde{M}U} \in [\lambda_{\min}, \lambda_{\max}],$$

$\lambda_{\min}, \lambda_{\max}$ are the extreme eigenvalues of $\widetilde{M}^{-1}M$, the proof is complete. ■

Corollary 1. The spectral radius of $I - B^{-1}A$ is bounded by the spectral radius of $I - \widetilde{M}^{-1}M$ and $\mathrm{cond}(B^{-1}A) \leq \mathrm{cond}(\widetilde{M}^{-1}M)$.

Proof. Follows directly from Theorem 1. ■

Corollary 2. The generalized conjugate gradient method from Axelsson [2] applied to the preconditioned system

$$B^{-1}A\begin{pmatrix}\mathrm{U}\\ P\end{pmatrix} = B^{-1}\begin{pmatrix}0\\ -F\end{pmatrix},$$

will have a convergence factor bounded by $(\sqrt{\varkappa}-1)/(\sqrt{\varkappa}+1)$, where $\varkappa = \mathrm{cond}(B^{-1}A) \leq \mathrm{cond}(\widetilde{M}^{-1}M)$.

Proof. Follows directly from [2] and corollary 1. ■

Let us consider the stationary iteration method

$$BX^{(k+1)} = \Phi + (B-A)X^{(k)} \tag{19}$$

where

$$X^{(k)} = \begin{pmatrix}\mathrm{U}^{(k)}\\ P^{(k)}\end{pmatrix}, \quad \Phi = \begin{pmatrix}0\\ -F\end{pmatrix} \quad \text{and } X^{(0)} \text{ is given.}$$

It is well known that the convergence rate of this algorithm is bounded by the spectral radius of the matrix $I-B^{-1}A$, denoted by $\rho(I-B^{-1}A)$. The Corollary 1 tells us that to increase the convergence speed of the iterations, it suffices to make $\rho(I-\widetilde{M}^{-1}M)$ as small as possible. Taking into account the construction of the matrices M and $\widetilde{M}$, we get

$$\begin{aligned}\rho(I-\widetilde{M}^{-1}M) &= \rho\left(I - \begin{pmatrix}\widetilde{M}^{-1} & 0\\ 0 & \widetilde{M}_{22}^{-1}\end{pmatrix}\begin{pmatrix}M_{11} & 0\\ 0 & M_{22}\end{pmatrix}\right)\\ &= \max\{\rho(I-\widetilde{M}_{11}^{-1}M_{11}), \rho(I-\widetilde{M}_{22}^{-1}M_{22})\}.\end{aligned} \tag{20}$$

$$\rho(I-\widetilde{M}_{11}^{-1}M_1) = \rho(I-\widetilde{M}_{11}^{-1/2}M_{11}\widetilde{M}_{11}^{-1/2})$$

$$(21) \qquad = \max_{U_1\neq 0}\left|\frac{U_1^T(I-\widetilde{M}_{11}^{-1/2}M_{11}\widetilde{M}_{11}^{1/2})U_1}{U_1^TU_1}\right|$$

$$= \max_{U_1\neq 0}\left|1-\frac{U_1^TM_{11}U_1}{U_1^T\widetilde{M}_{11}U_1}\right|$$

and similarly,

$$(22) \qquad \rho(I-\widetilde{M}_{22}^{-1}M_{22}) = \max_{U_2\neq 0}\left|1-\frac{U_2^TM_{22}U_2}{U_2^T\widetilde{M}_{22}U_2}\right|.$$

We start now the construction of the preconditioning matrix based on the local assembly procedure by optimal scaling of the lumped element mass matrices.

Let us denote by U_i^e, $i = 1, 2$ the vector of the values of the velocity function $U_i(x)$ at the points of the element $e \in T_h$. Then,

$$(23) \qquad a_e(U,U) \stackrel{\text{def}}{\equiv} \sum_{i=1}^{2}\int_e \frac{1}{k}U_i^2\,dx = \sum_{i=1}^{2} U_i^{e^T} M_{ii}^e U_i^e.$$

Here, M_{ii}^e are the element mass matrices. The global mass matrix is an assembly of M_{ii}^e over all $e \in T_h$, i.e.

$$\mathbf{U}^T M\mathbf{U} = \sum_{e\in T_h} a_e(U,U).$$

The lumped mass approximation $\widehat{M}$ to M is obtained, when instead of exact evaluation of the integrals in (23), one uses a trapezoidal rule in x_1 and a midpoint rule in x_2. Let $x = (x_1, x_2)$ be the center of the finite element $e \in T_h$. Then by definition

$$(24) \qquad U_1^{e^T}\widehat{M}_{11}^e U_1^e \equiv \frac{h^2}{2k}U_1^2\Big|_{(x_1-1/2h,\,x_2)} + \frac{h^2}{2k}U_1^2\Big|_{(x_1+1/2h,\,x_2)},$$

$$(25) \qquad U_2^{e^T}\widehat{M}_{22}^e U_2^e \equiv \frac{h^2}{2k}U_2^2\Big|_{(x_1,\,x_2-1/2h)} + \frac{h^2}{2k}U_2^2\Big|_{(x_1,\,x_2+1/2h)}.$$

Then the matrix $\widetilde{M}$ is defined by scaling the matrix $\widehat{M}$ element by element as follows:

$$(26) \qquad U^T\widetilde{M}U = \sum_{e\in T_h}(\omega_1^e U_1^{e^T}\widehat{M}_{11}^e U_1^e + \omega_2^e U_2^{e^T}\widehat{M}_{22}^e U_e^e);$$

i.e. $\widetilde{M}_{11}^e = \omega_1^e\widehat{M}_{11}^e$ and $\widetilde{M}_{22}^e = \omega_2^e\widehat{M}_{22}^e$ where the scaling parameters ω_1^e and ω_2^e have to be chosen in such way that $\rho(I-\widetilde{M}^{-1}M)$ is minimal.

Now, we show how to choose these parameters. In order to simplify our notations, we consider the choice of ω_1^e and denote M_{11}^e by M_e, $\widehat{M}_{11}^e$ by $\widehat{M}_e$, $\widetilde{M}_{11}^e$ by $\widetilde{M}_e$, U_1^e by U_e, and ω_1^e by ω_e.

Then by (24),

$$\widehat{M}_e = \begin{pmatrix} \widehat{m}_{11} & 0 \\ 0 & \widehat{m}_{22} \end{pmatrix}, \quad \widehat{m}_{11} = \frac{h^2}{2k(x_1-1/2h, x_2)}, \tag{27}$$
$$\widehat{m}_{22} = \frac{h^2}{2k(x_1+1/2h, x_2)},$$

and by (23),

$$M_e = \begin{pmatrix} m_{11} & m_{12} \\ m_{21} & m_{22} \end{pmatrix}.$$

Now taking into account the presentation of M_{11} by (23) and $\widetilde{M}_{11}$ by (26) (remember that below M_{11}^e is denoted by M_e and $\widetilde{M}_{11}^e$ by $\widetilde{M}_e$), we have

$$\left|1 - \frac{U_1^T M_{11} U_1}{U_1^T \widetilde{M}_{11} U_1}\right| = \left|1 - \frac{\sum_e U_e^T M_e U_e}{\sum_e U_e^T \widetilde{M}_e U_e}\right| \leq \max_{e \in T_h} \left|1 - \frac{U_e^T M_e U_e}{U_e^T \widetilde{M}_e U_e}\right|. \tag{28}$$

Combining (21) and (28), we conclude that

$$\rho(I - \widetilde{M}_{11}^{-1} M_{11}) \leq \max_{e \in T_h} \rho(I - \widetilde{M}_e^{-1} M_e). \tag{29}$$

From (29), we see that if we choose the local parameter ω_e such that $\rho(I-\widetilde{M}_e^{-1}M_e)$ reaches minimum, then $\rho(I-\widetilde{M}_{11}^{-1}M_{11})$ will also be near its minimum. We can use the same technique to optimize the spectral radius $\rho(I-\widetilde{M}_{22}^{-1}M_{22})$.

Thus we have the following result.

Theorem 2. Let $0 < \lambda_0^e < \lambda_1^e$ be the eigenvalues of matrix $\widehat{M}_e^{-1}M_e$ for each element e. If we choose local parameter ω_e on the element e to be

$$\lambda_e = \frac{\lambda_0^e + \lambda_1^e}{2}, \tag{30}$$

then the iteration method (19) is convergent and has a convergence rate independent of the mesh size h. The convergence speed is dominated by the inequality

$$\rho(I - B^{-1}A) \leq \max_e \frac{\lambda_1^e - \lambda_0^e}{\lambda_1^e + \lambda_0^e}. \tag{31}$$

Proof. By the assumption, $\widehat{M}_e^{-1}M_e$ has eigenvalues λ_0^e and λ_1^e, then, the matrix $I-\widetilde{M}_e^{-1}M_e = I-\omega_e^{-1}\widehat{M}_e^{-1}M_e$ has eigenvalues

$$1 - \frac{\lambda_0^e}{\omega_e} \quad \text{and} \quad 1 - \frac{\lambda_1^e}{\omega_e},$$

so

$$\rho(I-\widehat{M}_e^{-1}M_e) = \max\left\{\left|1-\frac{\lambda_0^e}{\omega_e}\right|, \left|1-\frac{\lambda_1^e}{\omega_e}\right|\right\}.$$

When $\omega_e = \frac{\lambda_1^e+\lambda_0^e}{2}$, we have

$$\rho(I-\widetilde{M}_e^{-1}M_e)\Big|_{\omega_e = \frac{\lambda_1^e+\lambda_0^e}{2}} = \frac{\lambda_1^e - \lambda_0^e}{\lambda_1^e + \lambda_0^e} < 1. \tag{32}$$

Now we show that, for any other values,

$$\rho(I-\widetilde{M}_e^{-1}M_e) \geq \frac{\lambda_1^e - \lambda_0^e}{\lambda_1^e + \lambda_0^e}.$$

If not, there exists an $\omega > 0$ such that

$$\left|1-\frac{\lambda_0^e}{\omega}\right| < \frac{\lambda_1^e - \lambda_0^e}{\lambda_1^e + \lambda_0^e} \tag{33}$$

and

$$\left|1-\frac{\lambda_1^e}{\omega}\right| < \frac{\lambda_1^e - \lambda_0^e}{\lambda_1^e + \lambda_0^e}. \tag{34}$$

Equation (33) implies

$$\begin{gathered} -\frac{\lambda_1^e - \lambda_0^e}{\lambda_1^e + \lambda_0^e} < 1 - \frac{\lambda_0^e}{\omega} < \frac{\lambda_1^e - \lambda_0^e}{\lambda_1^e + \lambda_0^e} \\ \frac{2}{\lambda_1^e + \lambda_0^e} < \frac{1}{\omega} < \frac{2\lambda_1^e}{(\lambda_1^e + \lambda_0^e)\lambda_0^e}. \end{gathered} \tag{35}$$

Similarly, (34) implies

$$\frac{2\lambda_0^e}{\lambda_1^e(\lambda_1^e + \lambda_0^e)} < \frac{1}{\omega} < \frac{2}{\lambda_1^e + \lambda_0^e}. \tag{36}$$

From (35) and (36), we get

$$\frac{\lambda_1^e + \lambda_0^e}{2} < \omega < \frac{\lambda_1^e + \lambda_0^e}{2},$$

which is a contradiction. So (29) and (32) imply that (31) holds; thus, (19) is convergent.

Finally, we show that the convergence rate is independent of the mesh size h. It is easy to show that λ_0^e and λ_1^e are the roots of the characteristic equation (M_e are symmetric),

$$\lambda^2 - \left(\frac{m_{11}}{\widehat{m}_{11}} + \frac{m_{22}}{\widehat{m}_{22}}\right)\lambda + \frac{m_{11}m_{22}}{\widehat{m}_{11}\widehat{m}_{22}} - \frac{m_{12}^2}{\widehat{m}_{11}\widehat{m}_{22}} = 0,$$

which are

$$(37)\qquad \begin{cases} \lambda_0^e = \dfrac{1}{2}\left[\dfrac{m_{11}}{\widehat{m}_{11}} + \dfrac{m_{22}}{\widehat{m}_{22}} - \left(\left(\dfrac{m_{11}}{\widehat{m}_{11}} + \dfrac{m_{22}}{\widehat{m}_{22}}\right)^2 + 4\dfrac{m_{12}^2}{\widehat{m}_{11}\widehat{m}_{22}}\right)^{\frac{1}{2}}\right], \\ \lambda_1^e = \dfrac{1}{2}\left[\dfrac{m_{11}}{\widehat{m}_{11}} + \dfrac{m_{22}}{\widehat{m}_{22}} + \left(\left(\dfrac{m_{11}}{\widehat{m}_{11}} + \dfrac{m_{22}}{\widehat{m}_{22}}\right)^2 + 4\dfrac{m_{12}^2}{\widehat{m}_{11}\widehat{m}_{22}}\right)^{\frac{1}{2}}\right] \end{cases}$$

so, (30), (31), and (37) imply that

$$\omega_e = \frac{1}{2}\left(\frac{m_{11}}{\widehat{m}_{11}} + \frac{m_{22}}{\widehat{m}_{22}}\right)$$

and

$$(38)\qquad \rho(I - \widetilde{M}_e^{-1}M_e) = \frac{\left(\left(\frac{m_{11}}{\widehat{m}_{11}} - \frac{m_{22}}{\widehat{m}_{22}}\right)^2 + 4\frac{m_{12}^2}{\widehat{m}_{11}\widehat{m}_{22}}\right)^{1/2}}{\left(\frac{m_{11}}{\widehat{m}_{11}} + \frac{m_{22}}{\widehat{m}_{22}}\right)}.$$

If we use the Simpson quadrature rule for evaluating the integrals in (23), we get the following approximate values for the elements of matrix M_e (for $e \in T_h$ with center $x = (x_1, x_2)$), which for smooth $k(x)$ are very close to the exactly computed ones:

$$(39)\qquad \begin{aligned} m_{11} &= \frac{h^2}{6}\left(\frac{1}{k(x_1 - 1/2h, x_2)} + \frac{1}{k(x_1, x_2)}\right) \\ m_{22} &= \frac{h^2}{6}\left(\frac{1}{k(x_1, x_2)} + \frac{1}{k(x_1 + 1/2h, x_2)}\right) \end{aligned}$$

$$m_{12} = m_{21} = \frac{h^2}{6}\frac{1}{k(x_1, x_2)}.$$

Substituting (39) into (38) and taking into account the definition of $\widehat{m}_{11}$ and $\widehat{m}_{22}$ by (27), we get

$$(40)\qquad \begin{aligned} \rho(I - \widetilde{M}_e^{-1}M_e) &= \frac{k(x_1 - 1/2h, x_2) + k(x_1 + 1/2h, x_2)}{k_1(x_1 - 1/2h, x_2) + 2k(x_1, x_2) + k(x_1 + 1/2h, x_2)} \\ &\le \frac{d}{1 + d}, \end{aligned}$$

where $d = \alpha_1/\alpha_0$ and α_0 and α_1 are defined in (6). Thus we have from (31) and (40)

$$\rho(I - B^{-1}A) \le \frac{d}{1 + d} < 1.$$

Here d is independent of the mesh size h and therefore, the algorithm (19) has convergence rate independent of h. ∎

We mention that the estimate (40) is not accurate enough for most cases. As a matter of fact, if we assume the coefficient $a(x)$ to be continuous on $\overline{\Omega} = [0,1]^2$, then on each element, we have the same relation as (40) with $d^e = \alpha_1^e/\alpha_0^e$, where α_1^e and α_0^e are upper and lower bounds of $k(x)$ on element e; i.e., $\alpha_0^e \leq k(x) \leq \alpha_1^e$, $x \in e$. Then

$$\rho(I - \widehat{M}_e^{-1} M_e) \leq \frac{d^e}{1+d^e} \tag{41}$$

and

$$\left|\frac{d^e}{1+d^e} - \frac{1}{2}\right| = \left|\frac{2d^e - 1 - d^e}{2(1+d^e)}\right| \leq \frac{1}{2}|d^e - 1| \leq \frac{1}{2\alpha_0^e}|\alpha_1^e - \alpha_0^e| \leq \frac{1}{2\alpha_0}|\alpha_1^e - \alpha_0^e|. \tag{42}$$

Since k is continuous on $\overline{\Omega}$, it is uniformly continuous. Thus $|\alpha_1^e - \alpha_0^e|$ tends to 0 when h approaches 0.

Corollary 3. Under the assumption of Theorem 2 and assuming coefficients $k(x)$ is continuous within each element, then the convergence rate of (19) tends to $\frac{1}{2}$ as h tends to 0.

In the case of discontinuous coefficients, there is no general conclusion except the estimate in (40). However, the same numerical experiment has been made to test the convergence property for discontinuous coefficients. We assume the function $k(x)$ has a jump along the diagonal line of the square domain $[0,1]^2$. The results suggest that the treatment of integration of discontinuous elements should be done very carefully. We cannot simply use the quadrature rule to obtain integral values of entries of the matrices M_e and $\widehat{M}_e$. Actually, we use a quadrature rule separately for any element e where discontinuity of the coefficient $k(x)$ takes place by partitioning the element into two parts (the discontinuity is then on the interface boundary) and then sum the two terms.

Finally, it should be mentioned that the construction of a "lumped mass" matrix $\widehat{M}$, described above, is somewhat complicated. Sometimes, the diagonal part of M could be used as $\widetilde{M}$.

5. Numerical Results and Conclusions

Some numerical experiments have been performed for the model equation

$$\begin{aligned} -\nabla \cdot (k \nabla p &= 0 \quad \text{in } (0,1)^2, \\ p &= 0 \quad \text{on the boundary of } (0,1)^2. \end{aligned}$$

To see that the convergence of the scheme is independent of the mesh size h, $h \to 0$ and the coefficient $k = k(x,y)$, we compute the numerical values U_1, U_2 and P with different mesh size h and different coefficients $k(x,y)$. The coefficients are given below:

$$k_I(x,y) = 1,$$
$$k_{II}(x,y) = e^{-x-y},$$
$$k_{III}(x,y) = \begin{cases} 1, & x > y, \\ 0.1, & x < y, \end{cases}$$
$$k_{IV}(x,y) = \begin{cases} 1, & x > y, \\ 0.01, & x < y, \end{cases}$$
$$k_V(x,y) = \begin{cases} e^{-x-y}, & x > y, \\ 0.1e^{-x-y}, & x < y. \end{cases}$$

The mesh sizes are $h = \frac{1}{2^m}$, $m = 4,5,6,7,8$. As a solution method we use the stationary iterative method (15) with preconditioning matrix B constructed by optimally scaled lumped element mass matrices discribed in the preceding section and $\widetilde{D} = 0$. The convergence factors are given in Table 1.

From these numerical results we can conclude that the convergence factor of the method is independent of the grid parameter and the coefficient $k(x,y)$ and they confirm the theoretical estimate given by Corollary 3. Comparing the results with those obtained in [11], where the convergence factors for the same test examples were estimated in some of the cases by 0.9, we see that the local element scaling is much more efficient than the global scaling.

TABLE

	Mesh size				
Coefficient	$1/2^4$	$1/2^5$	$1/2^6$	$1/2^7$	$1/2^8$
$k_I(x,y)$	0.493365	0.498830	0.499300	0.499535	0.499884
$k_{II}(x,y)$	0.496584	0.499463	0.498831	0.499686	0.499930
$k_{III}(x,y)$	0.494807	0.498198	0.499093	0.499761	0.499944
$k_{IV}(x,y)$	0.494658	0.497980	0.499249	0.499785	0.4999946
$k_V(x,y)$	0.493854	0.497781	0.498889	0.499860	0.0499971

Acknowledgements

The research reported here was partly supported by the Institute for Scientific Computation, University of Wyoming, Laramie, USA and in part by the Committee of Science, Bulgaria under Grant # 55-26-3-87.

References

[1] O. Axelsson, A survey of vectorizable preconditioning methods for large scale finite element matrix problems, Colloquium Topics in Applied Numerical Analysis, J.G. Verwer, ed., CWI Syllabus 4, Math. Center, Amsterdam, 1984.

[2] O. Axelsson. A generalized conjugate gradient, least square method, Numer. Math., 51 (1987), 209-227.

[3] O. Axelsson, P.S. Vassilevski. Algebraic multilevel preconditioning methods, I, Numer. Math. 56 (1989), 157-177.

[4] O. Axelsson, P.S. Vassilevski. A black box generalized conjugate gradient solver with inner iterations and variable step preconditioning, Report 9833, Department of Mathematics, University of Nijmegen, The Netherlands, 1989.

[5] R.E. Bank, B.D. Welfert and H. Yserentant. A class of iterative methods for solving saddle point problems, Numer. Math., 56 (1990), 645-666.

[6] J.H. Bramble and J.E. Pasciak. A preconditioning technique for indefinite systems resulting from mixed approximations of elliptic problems. Math. Comp., 181 (1988), 1-17.

[7] J.H. Bramble, J.E. Pasciak and J. Xu. The analysis of multigrid algorithms with non-nested spaces or non-inherited quadratic forms, Math. Comp. (to appear).

[8] R.E. Ewing and M.F. Wheeler, Computational aspects of mixed finite element methods, Numerical Methods for Scientific Computing, R.S. Stepleman, ed., Amsterdam, North Holland, 1983, 163-172.

[9] R.S. Falk and J.E. Osborn. Error estimates for mixed methods, RAIRO Numer. Anal. 14 (1980), 249-277.

[10] V. Girault and P.A. Raviart. Finite element approximation of the Navier-Stokes equations, Lecture Notes in Math. 749, Springer-Verlag, New York, 1981.

[11] P. Lu, M.B. Allen and R.E. Ewing. A semi-implicit iteration scheme using the multigrid method for mixed finite element equations, (submitted).

[12] P.A. Raviart and J.M. Thomas. A mixed finite element method for 2nd order elliptic problems, Mathematical Aspects of Finite Element Methods, Lecture Notes in Mathematics, 606, Galligani and E. Magenes, eds., Springer-Verlag, Berlin, 1977, 292-315.

A CLASS OF PRECONDITIONED CONJUGATE GRADIENT METHODS APPLIED TO FINITE ELEMENT EQUATIONS

IVAR GUSTAFSSON

Department of Computer Science, Chalmers University of Technology

S-412 96 Göteborg, Sweden

Abstract. It is well known that certain conditions have to be fulfilled to ensure existence of stable incomplete (IC) factorizations of a matrix.

Under additional conditions it is possible to construct *modified* incomplete (MIC) factorizations such that the order of convergence of the corresponding preconditioned conjugate gradient method is increased. For finite element discretizations of partial differential equations leading to matrices meeting these conditions, the number of iterations is of order $O(h^{-1/2})$, $h \to 0$, compared to $O(h^{-1})$ for the standard incomplete (IC) factorizations. Here, h is a mesh-parameter.

Since particularly one of the conditions, the so called MIC-condition, sometimes is violated in applications, while the IC-conditions are fulfilled, this point of the theory will be stressed. The simplest method, the so called MIC(0)* -method, will be used as a model in this context.

In this survey, methods for satisfying the fundamental conditions, when they are not primarily fulfilled, will be discussed. Two such ideas are spectral equivalence and hierarchical finite element techniques. A number of applications, e g from plane strain linear elasticity problems, where these ideas are used, will be presented.

For certain model-problems, it is possible to construct modified incomplete factorizations based only on element-matrix factors, such that the actual increase of the rate of convergence is still obtained. This idea will be briefly discussed too.

Key Words. Finite element equations, preconditioned iterative methods, conjugate gradients, incomplete factorization

1. Introduction. The process of discretizing a linear partial differential equation by a finite element method is illustrated by the following scheme:

$$
\begin{array}{ccccccc}
 & & (MIN) & \rightarrow & (MIN)_h & \Leftrightarrow & (QUAD) \\
 & & \Updownarrow & & & & \Updownarrow \\
(PDE) & \Leftrightarrow & (VAR) & \rightarrow & (VAR)_h & \Leftrightarrow & (LIN) \\
 & (1) & & (2) & & (3) &
\end{array}
$$

In step (1) the equation is given a suitable variational formulation. The equivalence holds under certain regularity requirements on the solution. In step (2) the problem is discretized by a finite element method. By an appropriate numbering of nodes in step (3) the system of linear equations *(LIN)* can be given a suitable form. The equivalence with a minimization problem is valid for symmetric (self-adjoint) problems. In this case, solving the system of linear equations is equivalent to minimizing a quadratic function. In order to solve such minimization problems we intend to use preconditioned conjugate gradient methods. We are particularly interested in the effect of the steps (1), (2) and (3) on the arising linear system of equations. It is well known that certain conditions have to be fulfilled in order for a system to be well suited for a preconditioned iterative method, see [11].

In section 2 we give a brief introduction to MIC-methods, including conditions for existence and fast rate of convergence. In section 3 we review some techniques in the three steps of the above scheme in order to get an easily solvable linear system. Our goal is to obtain fast methods, which are stable for rounding errors and do not require too much memory. Furthermore, they should be well suited for implementation on super and parallel computer architectures. In section 4 the ideas are applied to a linear elasticity problem and in section 5 an element-by-element version is introduced.

2. Modified Incomplete Factorizations. In the symmetric case a modified incomplete factorization of the matrix A can be stated as:

$$C = (L + D) D^{-1} (L^t + D) = A + \tilde{D} + R = \tilde{A} + R \tag{2.1}$$

$$Re = 0 \ , e = (1, 1, \cdots, 1)^t . \tag{2.2}$$

Here C is the preconditioning matrix, L is lower triangular, D and $\tilde{D}$ are positive diagonal matrices and R is the so called defect-matrix. L, D and $\tilde{D}$ are determined by the method. $\tilde{D}$ contains small perturbation entries and is used to ensure good convergence behavior. The row-sum criterion (2.2) is achieved by moving to the diagonal, entries that would have been dropped in a corresponding IC-factorization.

We define certain matrix-positions by

$$P^* = \{(i,j) ; a_{ij} \neq 0\}$$

$$P = \{(i,j) ; (i,j) \ P \Rightarrow l_{ij} = 0\}.$$

Clearly, P corresponds to positions where we allow fill-in during the incomplete factorization. The simplest method - MIC(0) - is defined by $P = P^*$. A strategy for obtaining successively more accurate factorizations reads as follow:

1. Start with $P = P^*$, that is the MIC(0)-method

2. Form $C = (L + D) D^{-1} (L^t + D)$ and $R = C - \tilde{A}$

3. Update $P = P \cup \{(i,j) ; r_{ij} \neq 0\}$

4. Repeat from 2 (if not satisfied with the current degree of accuracy).

Besides methods defined by this strategy, we also consider the so called MIC(0)*-method, which has the same sparsity pattern as the MIC(0)-method, but with no corrections to offdiagonal entries allowed during the factorization. These corrections are moved to the diagonal in order to preserve the row-sum criterion (2.2). This method can be viewed upon as a generalized SSOR-method i.e.

$$C = (L + D) D^{-1} (L^t + D)$$

with L being the lower triangular part of A.

For completeness we give the main results regarding existence - sometimes also called stability - and rate of convergence of MIC-methods. The proofs and further comments can be found in [11] and [12].

Stability. If A is a weakly diagonally dominant matrix then there exist MIC-factorizations (with $D > 0$) for any choice of P.

We note that if $\tilde{D} = 0$ then the same stability result holds for strictly diagonally dominant matrices.

Rate of convergence. For a second order PDE-problem on a mesh of size h the condition number $\kappa(A)$ of the stiffness matrix A is of order $O(h^{-2})$, $h \to 0$. Already for the least accurate modified incomplete factorization method, the MIC(0)*-method, the condition number $\kappa(C^{-1/2}AC^{-1/2})$ is of order $O(h^{-1})$ *if* the following conditions are all fulfilled:

(i) A is symmetric of order $N = O(h^{-n})$, $h \to 0$

(ii) A is a weakly diagonally dominant L-matrix. By definition all off-diagonal elements of an L-matrix are nonpositive.

(iii) The entries a_{ij} of A satisfy $a_{ii} \geq \sum_{j>i} a_{ij}(2 - m_i h)$, $i \in \mathcal{N}_1$,
where $m_i > 0$ are independent of h, $\mathcal{N}_1 \subseteq \mathcal{N} = \{i = 1, \cdots, N\}$ and the number of

indices in $\mathcal{N}_2 = \mathcal{N} \setminus \mathcal{N}_1$ is of order $O(h^{-n+1})$, $h \to 0$

(iv) The entries of $\tilde{D}$ satisfy $\tilde{d}_i = \begin{cases} \xi_1 h^2 a_{ii}, & i \in \mathcal{N}_1 \\ \xi_2 h\, a_{ii}, & i \in \mathcal{N}_2 . \end{cases}$

It should be observed that these conditions can easily be checked since they just include entries of the actual matrix A. The preconditioning parameters may be given the values $\xi_1 = \xi_2 = 0$ i.e. $\tilde{D} = 0$ for certain model problems see eg [8]. In order to get an optimal condition number, however, they should be positive. For instance for a model Laplace problem on the unit square, $\xi_{opt} = \pi^2/2$ and then $\kappa(A) \leq 2(1+\pi^{-1}h^{-1})$ for the MIC(0)-method. Actually, the same value ξ_{opt} is appropriate also for more accurate incomplete factorizations obtained by the strategy above.

The condition (i) is fulfilled for selfadjoint, elliptic PDE-problems in n space dimensions. The condition (ii) limits the choice of FEM-discretization to linear FEM over triangulation with angles $\leq \pi/2$ (or other simple elements like rectangles). In the next section we will discuss techniques to overcome this limitation.

The most crucial condition is (iii), the so called MIC-condition. It is fulfilled for natural ordering of mesh-points, such as row-wise or similar, but not for more general orderings like each-other-row-ordering or zig-zag-ordering. For the latter types of orderings an unmodified IC-factorization is usually preferable, see [11]. See however [2], where a red-black ordered system is reduced to a system, which is easy to solve by a MIC-method.

With a suitable ordering of nodes, the condition (iii) is fulfilled for a wide class of problems including Neumann-problems and problems with variable and even discontinuous material coefficients. For the appropriate definition of the preconditioning parameters in these cases we refer to [12].

3. Application to more general finite element discretizations. In order to meet the conditions (i) - (iv) of the previous section one has to be aware of the effects of the steps (1) - (3) in the introduction, to the resulting system of equations.

Choice of variational formulation. For instance, if the *Biharmonic problem*

$$\begin{cases} \Delta^2 u = f \\ \text{b.c.} \end{cases}$$

where b.c. stands for some kind of boundary conditions, is discretized by a standard finite difference or finite element method, the arising matrix is not an L-matrix. However, by using a mixed variable variational formulation in step (1), of the corresponding coupled system

$$\begin{cases} -\Delta u = w \\ -\Delta w = f \\ \text{b.c.} \end{cases}$$

a 2-by-2 block system arises, which can be reduced to a system suitable for efficient MIC-factorization techniques, for details see [13].

The resulting preconditioned conjugate gradient method converges in $O(h^{-1})$ iterations, compared to $O(h^{-2})$ for the unpreconditioned conjugate gradient method.

An additional advantage with the mixed variable variational formulation of the Biharmonic problem - a fourth order problem - is that simpler element-functions may be used. In fact, element-functions appropriate for second-order problems can be used here as well.

Mixed variable variational formulation can with great success also be used for *nonlinear diffusion problems*, see [5]. The effect of the nonlinearity can be reduced to just a diagonal matrix. This simplifies the updating during the iterative process. Furthermore, for discontinuous problems, this formulation gives in general a much more accurate approximation than the classical variational formulation.

Choice of discretization - Spectral equivalence. Although condition (ii) is not fulfilled for higher order FEM, this can be overcome by using the idea of spectral equivalence, see eg [12]. We assume that the mesh-refining is made in a uniform way giving a mesh characterized by the mesh-parameter h, for examples of such mesh-refinings see [4].

Let $A_h^{(p)}$ be the stiffness matrix for finite elements of degree p. Then $A_h^{(p)}$ and $A_h^{(1)}$ are *spectrally equivalent*, i.e. there exist constants m_0 and m_1, independent of h such that

$$m_0 \leq \frac{x^t A_h^{(p)} x}{x^t A_h^{(1)} x} \leq m_1 \, , \; \forall x \in R^N .$$

Here the matrices are of the same order N, i.e. the local element matrix $A_e^{(1)}$ is assembled over several linear elements (for instance p^2 triangles) making up an element corresponding to $A_e^{(p)}$.

The key for spectral equivalence is that the local element matrices $A_e^{(p)}, p = 1, \cdots$ have the same nullspace. For elements not meeting the boundary, the nullspace is spanned by the constant vector. The spectral equivalence constants $m_i, i = 1, 2$ can be calculated from local generalized eigen-problems

$$N_e^t A_e^{(1)} N_e v = \lambda N_e^t A_e^{(p)} N_e v ,$$

where N_e has linear independent columns not in the nullspace of $A_e^{(p)}$.

For an appropriate element division and ordering of nodes, $A_h^{(1)}$ can be made to satisfy the conditions (i) - (iv) and a MIC-factorization of this matrix can be used for $A_h^{(p)}$ as well. It is easily seen that the condition number of the preconditioned system is just increased by a factor m_1/m_0 (compared to the linear case) but it is still of order $O(h^{-1})$ for second order problems.

A drawback with this method is that we have to assemble (and store) both matrices $A_h^{(1)}$ and $A_h^{(p)}$.

Choice of discretization - Hierarchical elements. When using a hierarchical finite element basis with linear basis functions $\{\psi_i\}$ associated with vertex nodes and higher order basis functions $\{\phi_i\}$ associated with other nodes, we get a stiffness matrix in block-form:

$$\begin{bmatrix} B_h^{(p)} & F_h \\ F_h^t & A_h^{(1)} \end{bmatrix} \begin{bmatrix} u \\ v \end{bmatrix} = \begin{bmatrix} f \\ g \end{bmatrix}.$$

in which the vertex-nodes are numbered first. As the supscript indicates, $A_h^{(1)}$ is a stiffness matrix corresponding to linear finite elements i.e to $\{\psi_i\}$ and thus can be made to satisfy the conditions (i) - (iv) in section 2. Similarly, $B_h^{(p)}$ corresponds to $\{\phi_i\}$ and F_h to interaction of $\{\psi_i\}$ and $\{\phi_i\}$.

This method is studied in [4], particularly for the case $p = 2$ i.e. for quadratic approximation. It is shown that the matrix

$$\mathcal{A}_h = \begin{bmatrix} B_h^{(p)} & F_h \\ F_h^t & A_h^{(1)} \end{bmatrix}$$

is spectrally equivalent to

$$\mathcal{D}_h = \begin{bmatrix} B_h^{(p)} & 0 \\ 0 & A_h^{(1)} \end{bmatrix}$$

with the spectral equivalence bounds

$$1-\gamma \le \frac{x^t \mathcal{A}_h x}{x^t \mathcal{D}_h x} \le 1+\gamma \,, \ \forall x \in R^N.$$

Here, $0 < \gamma < 1$ is the constant in the strengthened Chauchy-Schwartz inequality

$$a(u,v) \le \gamma a(u,u)^{1/2} a(v,v)^{1/2} \quad , u \in U_h = span\{\phi_i\}\ , v \in V_h = span\{\psi_i\}$$

for the bounded, bilinear, coercive form $u(.,.)$ in the actual variational formulation.

We note that the constant vector $e \in U_h$, the only element in both U_h and V_h is the zero-element and that γ can be considered as the angle between the two spaces. In a similar way as for spectral equivalence, γ can be calculated from local generalized eigen-problems. For instance, the Laplace-problem on a rectangle with quadratic triangles and $p=2$ gives $\gamma=(2/3)^{1/2}$. For larger values of p or deformated triangles i.e. with some large angle, γ is closer to 1, see [14].

For solving the system A_h we can use a preconditioner for D_h at the cost of a factor $(1+\gamma)/(1-\gamma)$ in the condition number. Hence, it is preferable to keep γ away from 1.

Two choices of preconditioning matrices C_h for the system $\mathcal{A}_h$ are presented and studied in [4]. In the *diagonal-block* preconditioning

$$C_h = \begin{bmatrix} D & 0 \\ 0 & LL^t \end{bmatrix},$$

where LL^t is a MIC-factorization of $A_h^{(1)}$ and $D = \mathrm{diag}\,(B_h^{(p)})$. The reason why $B_h^{(p)}$ can be treated in this simple way is that its condition number is just of order $O(1)$, $h \to 0$. In the *full-block* preconditioning

$$C_h = \begin{bmatrix} L_2 & 0 \\ F_h L_2^{-t} & L_1 \end{bmatrix} \begin{bmatrix} L_2 & L_2^{-1} F_h^t \\ 0 & L_1^t \end{bmatrix},$$

where $L_1 L_1^t$ is a MIC-factorization of $A_h^{(1)}$ and $L_2 L_2^t$ is an IC-factorization of $B_h^{(p)}$.

For both methods it is observed that for accurate enough incomplete factorizations and for h not too small, $h \ge h_0$ say, the computational complexity is optimal i.e. proportional to the number of unknowns. Since the full-block factorization is more accurate, h_0 is smaller for this method for a given degree of accuracy of the (M)IC factorizations. On the other hand the full-block factorization method requires more operations per iteration than the simpler diagonal-block method so the latter method is still quite competitive.

The hierarchical finite element technique is generalized to a so called l-level method in [16] and to a recursive variant in [6] and [7].

Choice of node-numbering. It was mentioned already in the introduction that the numbering of nodes in step (3) is crucial for the MIC-condition. Sofisticated numberings such as red-black, zig-zag etc are not primarily suitable. It is sometimes possible, however, to reduce the derived block-system to systems of lower order, which satisfy the conditions. An early paper in this field is [2], where the reduction of problem-size is by a factor ½. Similarly, in a red-black-green ordering, the reduction is by a factor ¼. A somewhat different idea, affecting the step (3), is element-by-element factorization, see Section 5.

4. A linear elasticity problem. A linear, homogeneous, isotropic elasticity problem can be formulated in terms of *displacement vector* $u = (u_1, u_2, u_3)$, *stress tensor* $\sigma = \{\sigma_{ij}\}$ and *strain tensor* $\varepsilon = \{\varepsilon_{ij}\}$ satisfying

$$\varepsilon_{ij} = ½(\frac{\partial u_i}{\partial x_j} + \frac{\partial u_j}{\partial x_i})$$

in a domain Ω in R^3 with boundary Γ.

Assuming that the body is fixed along the part Γ_1 of the boundary and loaded by *body forces* $f = (f_1, f_2, f_3)$ and *boundary forces* $g = (g_1, g_2, g_3)$ the following equations and boundary conditions are obtained:

$$div(\sigma) + f = 0 \ in \ \Omega \tag{4.1a}$$

$$\sigma = \lambda \, div(u) I + 2\mu \, \varepsilon(u) \tag{4.1b}$$

$$u = u_0 \ on \ \Gamma_1 \tag{4.1c}$$

$$\sigma \bullet n = g \ on \ \Gamma_2 \tag{4.1d}$$

The material coefficients λ and μ are related to Poissons ratio ν and Young's modulus E through

$$\lambda = \frac{\nu E}{(1+\nu)(1-2\nu)}, \quad \mu = \frac{E}{2(1+\nu)}.$$

Furthermore, (4.1a) are equilibrium equations and (4.1b) are constitutive equations. Since $div(v) = trace \ \varepsilon(v)$, these equations can be written

$$-\lambda \, \Delta u - (\lambda + \mu) \, grad(div(u)) = f.$$

It is readily seen that a variational formulation for this problem can be stated as follows:

(VAR): Find u such that $u - u_0 \in V$ and $a(u, v) = L(v)$, $\forall v \in V$,

where $V \equiv \{v \in [H^1(\Omega)]^3 \ ; \ v = 0 \ on \ \Gamma_1\}$, with H^1 the usual Sobolev space. a is a symmetric, bilinear, bounded, coercive form and L is linear, bounded defined by

$$a(u,v) = \int_\Omega \{\lambda\, div(u)\, div(v) + 2\mu \sum_{i=1}^{3}\sum_{j=1}^{3} \varepsilon_{ij}(u)\, \varepsilon_{ij}(v)\} d\Omega$$

$$L(v) = \sum_{i=1}^{3} \{\int_\Omega f_i v_i \, d\Omega + \int_{\Gamma_1} g_i v_i \, d\Gamma\}.$$

A linear finite element discretization is introduced by the finite dimensional space $V_h \equiv \{v \in V : v|_K \in [P_1(K)]^3, \forall K \in C_h\}$ where C_h is a partitioning (e.g. a triangulation) of Ω. A set of basis functions for V_h is given by $\{\phi_i^{(1)}, \phi_i^{(2)}, \phi_i^{(3)}\}$, where $\phi_i^{(1)} = (\phi_i, 0, 0)$, $\phi_i^{(2)} = (0, \phi_i, 0)$, $\phi_i^{(3)} = (0, 0, \phi_i)$ with usual linear basis functions $\{\phi_i\}$.

In the plain strain case, i.e. in two dimensions, this problem is studied in [1]. When the nodes corresponding to $\{\phi_i^{(1)}\}$ are numbered first and then the nodes corresponding to $\{\phi_i^{(2)}\}$, the assembled matrix takes the form

$$\mathcal{A} = \begin{bmatrix} A & C \\ C^t & B \end{bmatrix},$$

where C is indefinite and A and B are weakly diagonally dominant L-matrices if the usual conditions are fulfilled, for instance in a triangulation with angles $\le \pi/2$ and suitable ordering of nodes, see section 2.

For a band-width-minimizing numbering $\phi_1^{(1)}, \phi_1^{(2)}, \phi_2^{(1)}, \phi_2^{(2)}, \phi_3^{(1)} \cdots$, however, the matrix is not an L-matrix - not even for this restricted triangulation - so neither an IC-factorization nor a MIC-factorization is suitable.

In order to study preconditionings for the matrix $\mathcal{A}$ we let

$$C = \begin{bmatrix} A & 0 \\ 0 & B \end{bmatrix}$$

be the so called *separate displacement component* (SDC) part of $\mathcal{A}$. The SDC part corresponds to the bilinear form (in three space dimensions)

$$\tilde{a}(u,v) = \int_\Omega \{(\lambda + 2\mu) \sum_{i=1}^{3} \frac{\partial u_i}{\partial x_i}\frac{\partial v_i}{\partial x_i} + 2\mu \sum_{i \ne j} \frac{\partial u_i}{\partial x_j}\frac{\partial v_i}{\partial x_j}\} \, d\Omega.$$

In the plain strain case with $\Gamma_1 = \Gamma$ the following spectral equivalence relation is shown in [1]:

$$c_1 \tilde{a}(u,u) \le a(u,u) \le c_2 \tilde{a}(u,u)$$

where $c_1 = (1-2\nu)/(1-\nu)$, $c_2 = 2$. Hence, the matrices $\mathcal{A}$ and C are spectrally equivalent with

$$\kappa(C^{-1/2} \mathcal{A}\, C^{-1/2}) \le \frac{2(1-\nu)}{1-2\nu} = \frac{2}{1-\tilde{\nu}},$$

where $\tilde{\nu} = \frac{\nu}{1-\nu}$, a transformation such that $0 \le \tilde{\nu} \le 1$. In a similar way it is also shown that for the bilinear form corresponding to the operator $-\Delta$,

$$a_0(u,v) = \int_\Omega \nabla u \nabla v \, d\Omega,$$

the spectral equivalence bounds

$$\frac{1-\tilde{\nu}}{2} a_0(u,u) \le a(u,u) \le \frac{3+\tilde{\nu}}{2} a_0(u,u)$$

holds, which means that the matrices $\mathcal{A}$ and

$$C_0 = \begin{bmatrix} A_0 & 0 \\ 0 & B_0 \end{bmatrix}$$

are spectrally equivalent with

$$\kappa(C_0^{-1/2} \mathcal{A}\, C_0^{-1/2}) \le \frac{3+\tilde{\nu}}{1-\tilde{\nu}}.$$

Here, A_0 and B_0 are matrices corresponding to discretization of $-\Delta$, actually, for the present choice of basis, $A_0 = B_0$.

For solving the preconditioning system, the involved systems with A, B or $A_0 = B_0$ may be solved by MICCG (MIC plus Conjugate Gradients) methods. In fact, with an appropriate ordering of the basis functions, the MIC-factorizations of A and B become more and more accurate when $\tilde{\nu} \to 1$.

The results so far in this section are generalized to three dimensional problems, to unisotropic (orthotropic) materials and to more general boundary conditions in [10].

For higher order finite elements, the SDC method can be combined with the hierarchical element method. For instance with quadratic approximation, $p=2$, the matrices A and B in $\mathcal{A}$ have a 2-by-2 block structure

$$A = \begin{bmatrix} A_{11}^{(2)} & A_{12} \\ A_{21} & A_{22}^{(1)} \end{bmatrix}, \quad B = \begin{bmatrix} B_{11}^{(2)} & B_{12} \\ B_{21} & B_{22}^{(1)} \end{bmatrix}.$$

For $\mathcal{A}$ we then may use a preconditioning with

$$C = \begin{bmatrix} A_{11}^{(2)} & 0 & 0 & 0 \\ 0 & A_{22}^{(1)} & 0 & 0 \\ 0 & 0 & B_{11}^{(2)} & 0 \\ 0 & 0 & 0 & B_{22}^{(1)} \end{bmatrix}$$

and as a bound for the condition number obtain

$$\kappa(C^{-1/2}\mathcal{A}\, C^{-1/2}) \le \frac{2}{1-\tilde{v}} \, \frac{1+\gamma}{1-\gamma}.$$

A nice observation in [15] is that γ depends only slightly on $\tilde{v}$. It is clear from Section 2 and the discussion above that $A_{22}^{(1)}$ and $B_{22}^{(1)}$, the (small) stiffness matrices corresponding to linear finite elements, can be well approximated by MIC, particularly for $\tilde{v}$ close to 1, and that $A_{11}^{(2)}$ and $B_{11}^{(2)}$ can be approximated by their diagonals. In this way we obtain a method which is essentially of optimal order for problems with almost incompressible material i.e. for $\tilde{v}$ close to 1. This is remarkable since the condition number of the preconditioned system goes to infinity as $\tilde{v}$ goes to 1.

5. An element-by-element factorization method. In this type of methods, preconditionings are constructed from factorizations of element matrices $A_e \in R^{n_e \times n_e}$, $e = 1, \cdots, E$, where E is the total number of elements in the partitioning of Ω. Since these matrices are positive (semi-)definit, there exist factorizations of the form $A_e = L_e L_e^t$, for details see [14], or $A_e = (L_e + D_e) D_e^+ (L_e^t + D_e)$, for details see [16]. Here $D_e^+ \ge 0$ is the generalized inverse of D_e. For uniquely defined Boolean connectivity matrices $N_e \in R^{N \times n_e}$, the assembled matrix A (of order N) can be written $A = \sum_{e=1}^{E} N_e A_e N_e^t$.

The matrices L_e and D_e may now be assembled into

$$L = \sum_{e=1}^{E} N_e L_e N_e^t \ , \quad D = \sum_{e=1}^{E} N_e D_e N_e^t .$$

If the numbering of nodes is made such that increasing local node-numbers correspond to increasing global node-numbers, then L becomes lower triangular and as preconditioning we can use

$$C = LL^t \quad \text{or} \quad C = (L + D)D^{-1}(L^t + D), \text{ respectively.}$$

The assembled matrices do not have to be stored since the assembling may be done locally, temporarily when solving the triangular systems involved, see [14].

For a mild condition on the numbering of nodes, which is easily fulfilled in practice - see [16], $D > 0$ i.e. the factorization exists. Observe that restrictions like linear finite elements and small angles do not have to be asserted for this kind of factorization. For pure Neumann problems the second kind of preconditioning is applicable with the slight change

$$C = (L + D)D^{+}(L^t + D), \tag{5.1}$$

where D^{+} is the generalized inverse of D and all elements but the last one of D are positive.

A more general element-by-element factorization can be constructed by letting A_e be super-element matrices, i.e. assembled from some element matrices. In this way the degree of accuracy of the factorization can be varied.

Modified element-by-element factorizations can be derived in a way similar to the standard MIC-methods. For (5.1) the modification reads

$$C = (L + (1 + \xi h)D)(1 + \xi h)^{-1} D^{+}(L^t + (1 + \xi h)D)$$

and similarly for the other preconditionings. $\xi > 0$ is a preconditioning parameter and h is the usual mesh-parameter.

Existence of modified element-by-element factorizations follows in the same way as for unmodified. In computer tests, see [14], the rate of convergence has been found to be comparable to the MIC(0) *-method for smooth Dirichlet problems. For Neumann problems and discontinuous-material problems the modified element-by-element method seems to be less efficient than the MIC(0)*-method.

Besides easily fulfilled conditions for existence, a great advantage with the element-by-element method is the small amount of memory required. If the mesh is refined uniformly, the total storage requirement for the (modified) element-by-element factorization is of order $O(1)$, $h \to 0$, for details see [14].

6. Conclusions. From this survey on MIC-methods, it becomes clear that this type of methods is competitive for solving linear finite element equations from wide classes of applications. One has to be aware, however, of how to make suitable variational formulations, finite element discretizations and node numbering. The methods have a high rate of convergence and have small storage requirements - particularly the element-by-element method. Furthermore, by using the technique in [3] for calculating residuals, they can be made stable for rounding errors. Finally, methods of this class are well suited for implementation on newer computer architectures, see for instance [8] and [9].

REFERENCES

[1] O. AXELSSON and I. GUSTAFSSON, *Iterative Methods for the Solution of the Navier Equations of Elasticity*, Comput. Meths. Appl. Mech. Engrg. 15 (1978), 241-258.

[2] O. AXELSSON and I. GUSTAFSSON, *On the use of Preconditioned Conjugate Gradient Methods for Red-Black Ordered Five-point Difference Schemes*, J. Comp. Physics 35 (1980), 284-289.

[3] O. AXELSSON and I. GUSTAFSSON, *A Preconditioned Conjugate Gradient Method for Finite Element Equations, which is Stable for Rounding Errors*, Information Processing 80, ed: S. H. Lavington, North Holland Publishing Company (1980), 723-728.

[4] O. AXELSSON and I. GUSTAFSSON, *Preconditioning and Two-Level Multigrid Methods of Arbitrary Degree of Approximation*, Math. Comp. 40 (1983), 219-242.

[5] O. AXELSSON and I. GUSTAFSSON, *An Efficient Finite Element Method for Nonlinear Diffusion Problems*, Report 84.06 R, Department of Computer Sciences, Chalmers University of Technology, Göteborg, Sweden, 1984.

[6] O. AXELSSON and P. S. VASSILEVSKI: *Algebraic Multilevel Preconditioning Methods I*, Report 8811, Department of Mathematics, Catholic University, Nijmegen, Holland, 1988.

[7] O. AXELSSON and G. LINDSKOG, *A Recursive Two-Level Method for Boundary Value Problems Discretized by Quadratic Finite Elements*, Report 8, Numerical Analysis Group, Department of Computer Sciences, Chalmers University of Technology, Göteborg, Sweden, 1988.

[8] O. AXELSSON, *On Iterative Solution of Elliptic Difference Equations on a Mesh-Connected Array of Processors*, International Journal of High Speed Computing 1 (1989), 165-183.

[9] O. AXELSSON, G. CAREY and G. LINDSKOG, *On a Class of Preconditioned Iterative Methods for Parallel Computers*, Technical Report, Dept of Aerospace Eng. and Eng.

Mech. The University of Texas at Austin, USA, 1989.

[10] R. BLAHETA, *An Incomplete Factorization Preconditioning Technique for Solving Linear Elasticity Problems*, submitted to the special issue of BIT on PCG methods, 1989.

[11] I. GUSTAFSSON, *Stability and Rate of Convergence of Modified Incomplete Cholesky Factorization Methods, Report 79.02 R*, Department of Computer Sciences, Chalmers University of Technology, Göteborg, Sweden, 1979.

[12] I. GUSTAFSSON, *Modified Incomplete Cholesky (MIC) Factorizations*, Preconditioning Methods - Theory and Applications, ed: D. Evans, Gordon and Breach Publishers, New York - London - Paris (1983), 265-293.

[13] I. GUSTAFSSON, *A Preconditioned Iterative Method for the Solution of the Biharmonic Problem*, IMA Journal of Numerical Analysis 4 (1984), 55-67.

[14] I. GUSTAFSSON and G. LINDSKOG, *A Preconditioning Technique based on Element Matrix Factorizations*, Comput. Meths. Appl. Mech. Engrg. 55 (1986), 201-220.

[15] M. JUNG, U. LANGER and U. SEMMLER, *Preconditioned Conjugate Gradient Methods for Solving Linear Elasticity Finite Element Equations*, submitted to the special issue of BIT on PCG methods, 1989.

[16] E. F. KAASSCHIETER, *A General Finite Element Preconditioning for the Conjugate Gradient Method*, submitted to the special issue of BIT on PCG methods, 1989.

RECENT VECTORIZATION AND PARALLELIZATION OF ITPACKV *

DAVID R. KINCAID AND THOMAS C. OPPE †

Abstract. A status report is given on recent work on the vectorization and parallelization of the ITPACKV 2C [4] software. This package has been modified for improved performance on vector supercomputers resulting in a new version called ITPACKV 2D [10] that includes *wavefront* techniques. The main reason for these changes was to obtain improved vector performance of several methods that did not vectorize in the previous version. Results are also presented on initial efforts at modifying this package to produce a *parallel* ITPACKV suitable for execution on shared-memory parallel computers with relatively small numbers of processors.

Key words. iterative methods, large sparse systems of linear equations, numerical solution of partial differential equations, parallelization, polynomial acceleration procedures, research-oriented software, shared-memory parallel computers, supercomputers, vector computers, vectorization.

AMS(MOS) subject classifications. 65F10, 65M10, 68B99

1. Introduction. A number of research-oriented software packages for solving large sparse systems of linear algebraic equations have been developed within the Center for Numerical Analysis at The University of Texas at Austin. The ITPACK Project [9] was established over a decade ago for developing and evaluating iterative algorithms for solving large sparse systems of linear algebraic equations by writing research-oriented software based on theoretical research. The emphasis has been on linear systems arising in the solution of partial differential equations by finite-difference and finite-element discretizations. Several computer packages have been developed in connection with this research and used as software tools for carrying out experimental studies using various iterative algorithms over a wide range of problems. The ITPACK packages have been modified, improved, enhanced, and changed through various versions.

In the ITPACKV packages [7] [10], the basic methods Jacobi (J), Reduced System (RS), and Symmetric Successive Overrelaxation (SSOR) are combined with acceleration procedures—Chebyshev (Semi-Iteration, SI) and Conjugate Gradient (CG). Two major features of the package are automatic selection of the parameters in the acceleration procedures and the use of stopping tests that are both automatic and accurate. While the routines in the package can be called with any linear system containing positive diagonal elements, they are most successful in solving systems with symmetric positive definite or mildly nonsymmetric coefficient matrices.

A status report is given on recent work on the vectorization and parallelization of the ITPACK 2C[4] software. This package has been modified for improved performance on vector supercomputers resulting in a new version called ITPACKV 2D [10] that includes *wavefront* techniques. This allows the vectorization of several methods that were not vectorized in the previous version, *i.e.*, successive overrelaxation (SOR), symmetric SOR conjugate gradient (SSORCG), and symmetric SOR semi-iterative (SSORSI)—all using the natural ordering.

* Received by editors June 6, 1990.

† First author: Center for Numerical Analysis, The University of Texas at Austin, Austin, Texas 78713–8510 USA. Second author: Supercomputer Computations and Research Institute, Florida State University, Tallahassee, Florida 3206–4052 USA.

Results are also presented on initial efforts at modifying this package to produce a *parallel* ITPACK suitable for execution on shared-memory parallel computers with relatively small numbers of processors (*e.g.*, Alliant FX/8, Sequent Symmetry, etc.).

In Section 2, we begin with a review of the methods in ITPACKV. Then in Section 3 some brief background material on basic iterative methods is given. The polynomial acceleration procedures used in the package are discussed is Section 4. Techniques for the vectorization and parallelization of the software are presented in Sections 5 and 6, respectively. Finally, numerical results are given in Section 7 and conclusions are drawn in Section 8.

2. ITPACKV Package. ITPACKV is a collection of Fortran subroutines for solving large sparse linear systems by using accelerated iterative algorithms with adaptive determination of the acceleration parameters and automatic stopping procedures. The basic iterative procedures in the package are:

JCG	Jacobi Conjugate Gradient
JSI	Jacobi Semi-Iteration
SOR	Successive Overrelaxation
SSORCG	Symmetric SOR Conjugate Gradient
SSORSI	Symmetric SOR Semi-Iteration
RSCG	Reduced System Conjugate Gradient
RSSI	Reduced System Semi-Iteration

Two orderings of the unknowns in the linear system are available—the natural (lexicographical) ordering and the red-black (checker-board) ordering. The reduced system routines (RSCG and RSSI) require a 2×2 block partitioning of the coefficient matrix such as that obtained from the red-black ordering of the system. For additional details, see [4].

3. Basic Iterative Methods. We begin with some brief background material on iterative methods. Iterative methods can be utilized to solve linear systems

$$Ax = b, \tag{1}$$

where the coefficient matrix A is large and sparse. Here both A and b are given and we seek an efficient method for finding an acceptable value for x.

Suppose Q is a nonsingular matrix and the linear system (1) is *preconditioned* as follows

$$Q^{-1}Ax = Q^{-1}b.$$

Letting $G = I - Q^{-1}A$ and $k = Q^{-1}b$, an iterative method is suggested of the form

$$x^{(n+1)} = Gx^{(n)} + k.$$

Here G denotes the *iteration* matrix and Q is called the *preconditioning* matrix or the *splitting* matrix since A can be written as $A = Q - (Q - A)$. Such a preconditioning is done implicitly as part of the iterative procedure and not as shown here.

We begin by diagonally scaling the linear system (1) so that it has a unit diagonal. Assuming that the diagonal elements of A are nonzero, the following two-sided scaling accomplishes this

$$(D^{-\frac{1}{2}}AD^{-\frac{1}{2}})(D^{\frac{1}{2}}x) = (D^{-\frac{1}{2}}b),$$

using the diagonal matrix $D^{\frac{1}{2}} = \text{diag}(\sqrt{|a_{ii}|})$. Notice that if A is symmetric then the coefficient matrix in the scaled system preserves this property. If $a_{ii} < 0$, then the i-th equation in the system is multiplied by -1 before and after the solution procedure. Now we have an equivalent system to (1) but of the form

$$(I - L - U)\widehat{x} = \widehat{b},$$

where L is a strictly lower triangular matrix, U is a strictly upper triangular matrix, $\widehat{x} = D^{\frac{1}{2}}x$, and $\widehat{b} = D^{-\frac{1}{2}}b$. Clearly, upon convergence of the iterative process based on $\widehat{x}$, a postprocessing is required, *i.e.*, $x = D^{-\frac{1}{2}}\widehat{x}$. In addition during the algorithms based on the SOR and SSOR methods, this system is scaled by the relaxation factor ω each time the current estimate for ω is changed. This will be discussed more later.

For simplicity, we now drop the hats and assume that the linear system is of the form

$$(I - L - U)x = b, \tag{2}$$

where L is strictly lower triangular and U is strictly upper triangular.

If the red-black ordering is used, the linear system is permuted and assumes the special form

$$\begin{bmatrix} I & -F_R \\ -F_B & I \end{bmatrix} \begin{bmatrix} x_R \\ x_B \end{bmatrix} = \begin{bmatrix} b_R \\ b_B \end{bmatrix}, \tag{3}$$

where x_R and x_B are the *red* and *black* unknowns, respectively. Similarly, $-F_R$ and $-F_B$ are the corresponding coefficients for these unknowns with b_R and b_B the related right-hand sides. The off-diagonal blocks are usually not as sparse as the original system. However, the 2×2 block structure of this system can be used to advantage. Eliminating x_R results in a *reduced system* involving just the black unknowns

$$(I - F_B F_R)x_B = (F_B b_R + b_B). \tag{4}$$

Similarly, we can obtain another reduced system for the red unknowns

$$(I - F_R F_B)x_R = (F_R b_B + b_R).$$

Putting these two reduced systems together, we have the *double reduced system*

$$\begin{bmatrix} I - F_R F_B & 0 \\ 0 & I - F_B F_R \end{bmatrix} \begin{bmatrix} x_R \\ x_B \end{bmatrix} = \begin{bmatrix} F_R b_B + b_R \\ F_B b_R + b_B \end{bmatrix}. \tag{5}$$

We now briefly describe the basic iterative methods in this package. We are interested in identifying the operations of primary importance for vector and parallel computations. It will be shown that these are matrix-vector operations. Since the sum of two or three vectors is obviously a vectorizable/parallelizable operation, we will not mention them repeatedly.

Jacobi. The basic iterative method known as the Jacobi method can be written as

$$x^{(n+1)} = (L + U)\, x^{(n)} + b.$$

The iteration matrix is $G = L + U$ and the splitting matrix is $Q = I$. We see that the primary operation is a matrix-vector multiplication of the form $y = (L + U)x$.

Reduced System. The Jacobi method can be applied to the reduced system (4) and it would be written as

$$x_B^{(n+1)} = (F_B F_R) x_B^{(n)} + (F_B b_R + b_B).$$

This is the reduced system method with iteration matrix $G = F_B F_R$ and splitting matrix $Q = I$. Rather than forming the matrix $F_B F_R$, this iterative scheme can be accomplished in two-steps as

$$\begin{cases} x_R^{(n+1)} = F_R x_B^{(n)} + b_R \\ x_B^{(n+1)} = F_B x_R^{(n+1)} + b_B. \end{cases}$$

The primary operations are matrix-vector multiplications of the form $y = F_R x$ and $y = F_B x$.

Double Reduced System. Applying the Jacobi method to the double reduced system (5), we have

$$\begin{cases} x_R^{(n+1)} = F_R F_B x_R^{(n)} + F_R b_B + b_R \\ x_B^{(n+1)} = F_B F_R x_B^{(n)} + F_B b_R + b_B. \end{cases}$$

The iteration matrix is

$$G = \begin{bmatrix} F_R F_B & 0 \\ 0 & F_B F_R \end{bmatrix}.$$

This method would be implemented as a four-step procedure, and therefore, the primary operations would be matrix-vector multiplications of the form $y = F_R x$ and $y = F_B x$.

Successive Overrelaxation. For the SOR and SSOR basic iterative methods, there is an optimum relaxation parameter, say ω. The linear system (2) can be scaled using ω as follows

$$(\omega I - \widetilde{L} - \widetilde{U})x = \widetilde{b}, \tag{6}$$

where $\widetilde{L} = \omega L$, $\widetilde{U} = \omega U$, and $\widetilde{b} = \omega b$. Rewriting, we have

$$x = \widetilde{L}x + \widetilde{U}x + \widetilde{b} + (1 - \omega)x.$$

This suggests an iteration procedure and, since $\widetilde{L}$ is a strictly lower triangular matrix, we can use the new iteration values as soon as they become available. Thus, we have the SOR method

$$x^{(n+1)} = \widetilde{L}x^{(n+1)} + \widetilde{U}x^{(n)} + \widetilde{b} + (1 - \omega)x^{(n)}, \tag{7}$$

so that

$$(I-\widetilde{L})x^{(n+1)}=\widetilde{U}x^{(n)}+\widetilde{b}+(1-\omega)x^{(n)}.$$

Thus, the iteration matrix is

$$G=\mathcal{L}_\omega=(I-\widetilde{L})^{-1}(\widetilde{U}+(1-\omega)I).$$

Moreover, it can be shown that $\mathcal{L}_\omega=I-Q^{-1}A$ where $Q=(I-\widetilde{L})$ is the splitting matrix. The primary operations are a matrix-vector multiplication of the form $y=\widetilde{U}x$ and solving a lower triangular system of the form $(I-\widetilde{L})x=y$, which can be done using forward substitution.

For the red-black ordering, system (3) is scaled by ω to form

$$\begin{bmatrix} \omega I & -\widetilde{F}_R \\ -\widetilde{F}_B & \omega I \end{bmatrix}\begin{bmatrix} x_R \\ x_B \end{bmatrix}=\begin{bmatrix} \widetilde{b}_R \\ \widetilde{b}_B \end{bmatrix},$$

where $\widetilde{F}_R=\omega F_R$, $\widetilde{F}_B=\omega F_B$, $\widetilde{b}_R=\omega b_R$, and $\widetilde{b}_B=\omega b_B$. Comparing this system to system (6), we see that

$$\widetilde{L}=\begin{bmatrix} 0 & 0 \\ \widetilde{F}_B & 0 \end{bmatrix}, \quad \text{and} \quad \widetilde{U}=\begin{bmatrix} 0 & \widetilde{F}_R \\ 0 & 0 \end{bmatrix}.$$

Now it follows from (7) that the SOR method for the red-black system can be written as

$$\begin{cases} x_R^{(n+1)}=\widetilde{F}_R x_B^{(n)}+\widetilde{b}_R+(1-\omega)x_R^{(n)} \\ x_B^{(n+1)}=\widetilde{F}_B x_R^{(n+1)}+\widetilde{b}_B+(1-\omega)x_B^{(n)}. \end{cases}$$

In this case, the primary operations are two matrix-vector multiplications of the form $y=\widetilde{F}_R x$ and $y=\widetilde{F}_B x$. The iteration matrix is then

$$G=\begin{bmatrix} (1-\omega)I & \widetilde{F}_R \\ (1-\omega)\widetilde{F}_B & \widetilde{F}_B\widetilde{F}_R+(1-\omega)I \end{bmatrix},$$

and the splitting matrix is

$$Q=\begin{bmatrix} I & 0 \\ -\widetilde{F}_B & I \end{bmatrix}.$$

Gauss-Seidel. When $\omega=1$, the SOR method reduces to the well-known Gauss-Seidel method

$$(I-L)x^{(n+1)}=Ux^{(n)}+b.$$

with iteration matrix $G=\mathcal{L}=(I-L)^{-1}U$ and splitting matrix $Q=(I-L)$. The primary operations are of the form $y=Ux$ and solving $(I-L)x=y$.

When red-black ordering is used, the SOR method with $\omega=1$ reduces to the Gauss-Seidel method of the form

$$\begin{cases} x_R^{(n+1)}=F_R x_B^{(n)}+b_R \\ x_B^{(n+1)}=F_B x_R^{(n+1)}+b_B, \end{cases}$$

which is the reduced system method discussed previously.

Symmetric SOR. The SSOR basic iterative method is a two-step procedure. First, an application of a *forward* SOR iteration

$$(I - \tilde{L})x^{(n+\frac{1}{2})} = \tilde{U}x^{(n)} + \tilde{b} + (1-\omega)\,x^{(n)},$$

and then a *backward* SOR iteration

$$(I - \tilde{U})x^{(n+1)} = \tilde{L}x^{(n+\frac{1}{2})} + \tilde{b} + (1-\omega)\,x^{(n+\frac{1}{2})}.$$

Notice in the second-half iteration the roles of $\tilde{L}$ and $\tilde{U}$ have been reversed from the first-half. This corresponds to traversing the unknowns in the opposite order from the first part of the iterative sweep. In general, the relaxation factor ω for the SSOR method is not the same as that for the SOR method. The SSOR iteration matrix is

$$G = \mathcal{S}_\omega = \mathcal{U}_\omega \mathcal{L}_\omega,$$

where

$$\begin{cases} \mathcal{L}_\omega = (I-\tilde{L})^{-1}(\tilde{U} + (1-\omega)I) \\ \mathcal{U}_\omega = (I-\tilde{U})^{-1}(\tilde{L} + (1-\omega)I). \end{cases}$$

It can be shown that $\mathcal{S}_\omega = I - Q^{-1}A$ with $Q^{-1} = (2-\omega)(I-\tilde{U})^{-1}(I-\tilde{L})^{-1}$, and therefore, the splitting matrix is

$$Q = \frac{1}{2-\omega}(I-\tilde{L})(I-\tilde{U}).$$

Now the primary operations are two matrix-vector multiplications of the form $y = \tilde{U}x$ and $y = \tilde{L}x$, a forward solve of $(I-\tilde{L})x = y$, and a backward solve of $(I-\tilde{U})x = y$. The forward solve is just a forward substitution whereas the backward solve is a back substitution.

Symmetric Gauss-Seidel. When $\omega = 1$, the SSOR method reduces to the symmetric Gauss-Seidel method. Thus, we have a *forward* Gauss-Seidel sweep

$$(I-L)x^{(n+\frac{1}{2})} = Ux^{(n)} + b,$$

and then a *backward* Gauss-Seidel sweep

$$(I-U)x^{(n+1)} = Lx^{(n+\frac{1}{2})} + b.$$

Clearly, the iteration matrix is

$$G = \mathcal{U}\mathcal{L} = (I-U)^{-1}L(I-L)^{-1}U,$$

and the splitting matrix is

$$Q = (I-L)(I-U),$$

since $\mathcal{L} = I - Q^{-1}A$. The primary operations are matrix-vector multiplications of the form $y = Ux$ and $y = Lx$ plus forward and backward solves of systems of the form $(I-L)x = y$ and $(I-U)x = y$, respectively.

In the case of the red-black ordering, the optimal relaxation parameter for the SSOR method is $\omega = 1$ and it reduces to the symmetric Gauss-Seidel method. In this case, we have a *forward* Gauss-Seidel sweep

$$\begin{cases} x_R^{(n+\frac{1}{2})} = F_R x_B^{(n)} + b_R \\ x_B^{(n+\frac{1}{2})} = F_B x_R^{(n+\frac{1}{2})} + b_B, \end{cases}$$

and a *backward* Gauss-Seidel sweep

$$\begin{cases} x_B^{(n+1)} = F_B x_R^{(n+\frac{1}{2})} + b_B \\ x_R^{(n+1)} = F_R x_B^{(n+1)} + b_R. \end{cases}$$

Here the primary operations are four matrix-vector multiplications of the form $y = F_R x$ and $y = F_B x$. Compressing these equations, we have

$$\begin{cases} x_B^{(n+\frac{1}{2})} = F_B F_R x_B^{(n)} + F_B b_R + b_B \\ x_R^{(n+1)} = F_R F_B x_R^{(n+\frac{1}{2})} + F_R b_B + b_R, \end{cases}$$

which is the Gauss-Seidel method for the double reduced system (5).

Summary. In summary, the primary vectorizable and parallelizable operations in the iterative algorithms in **ITPACKV** are of the following forms:

- matrix-vector multiplications

$$y = \widetilde{L}x, \quad y = \widetilde{U}x, \quad y = \widetilde{F}_B x, \quad y = \widetilde{F}_R x,$$

- forward solves

$$(I - \widetilde{L})x = y,$$

- backward solves

$$(I - \widetilde{U})x = y.$$

4. Polynomial Accelerators in ITPACKV. There are two polynomial acceleration algorithms used in the **ITPACKV** package: three-term versions of Chebyshev and conjugate gradient acceleration.

Chebyshev Acceleration. Chebyshev acceleration can be written as

$$x^{(n+1)} = \rho_{n+1}\left\{\gamma\delta^{(n)} + x^{(n)}\right\} + (1 - \rho_{n+1})\, x^{(n-1)},$$

where

$$\delta^{(n)} = Gx^{(n)} + k - x^{(n)}.$$

Here $\delta^{(n)}$ is called the *pseudo-residual* vector and it can be computed recursively by

$$\delta^{(n+1)} = \rho_{n+1}\left\{\gamma G\delta^{(n)} + (1-\gamma)\delta^{(n)}\right\} + (1 - \rho_{n+1})\, \delta^{(n-1)}.$$

The primary operations are matrix-vector multiplications of the form $y \leftarrow Gx$ and linear combinations of vectors. The iterative parameters ρ_{n+1} and γ are computed based on estimates for the maximum eigenvalue $M(G)$ and the minimum eigenvalue $m(G)$ of the iteration matrix G. Let M_E and m_E be these estimates, respectively. Formulæ for the acceleration parameters are given by

$$\begin{cases} \gamma = \dfrac{2}{2 - M_E - m_E}, & \sigma_E = \dfrac{M_E - m_E}{2 - M_E - m_E}, \\ \rho_{n+1} = \left[1 - \left(\dfrac{\sigma_E}{2}\right)^2 \rho_n\right]^{-1}, & n \geq 2, \end{cases}$$

$\rho_1 = 1$, and $\rho_2 = 1/(1 - \sigma_E^2/2)$. These calculations are scalar in nature and consume a minimum amount of computer time as compared to matrix-vector operations. The package contains an adaptive procedure for improving these estimates, if good ones are not already available. Operations of this type are clearly not vectorizable or parallelizable. While they can slow the overall execution time down, they are not as time-consuming as the primary operations mentioned above.

Conjugate Gradient Acceleration. The formulæ for conjugate gradient acceleration are given by

$$x^{(n+1)} = \rho_{n+1}\left\{\gamma_{n+1}\delta^{(n)} + x^{(n)}\right\} + (1 - \rho_{n+1})\, x^{(n-1)},$$

where the pseudo-residual $\delta^{(n)} = Gx^{(n)} + k - x^{(n)}$ can be computed recursively by

$$\delta^{(n+1)} = \rho_{n+1}\left\{\gamma_{n+1}G\delta^{(n)} + (1 - \gamma_{n+1})\delta^{(n)}\right\} + (1 - \rho_{n+1})\, \delta^{(n-1)}.$$

As before, the primary operations are matrix-vector multiplications of the form $y \leftarrow Gx$ and linear combinations of vectors. Unlike Chebyshev acceleration, the computation of the parameters ρ_{n+1} and γ_{n+1} can be quite time consuming since they involve dot-products $\langle \cdot, \cdot \rangle$. However, no eigenvalue estimates are required, which can be a definite advantage in removing an unknown factor from the algorithm.

As with the previous acceleration procedure, the calculation of the iteration parameters are scalar in nature but they are directly dependent on the speed of the dot-product operation on the computer being used. The formulæ for the conjugate gradient acceleration parameters are given by

$$\begin{cases} \gamma_{n+1} = \left[1 - \dfrac{\langle W\delta^{(n)}, WG\delta^{(n)}\rangle}{\langle W\delta^{(n)}, W\delta^{(n)}\rangle}\right]^{-1}, \\ \rho_{n+1} = \left[1 - \left(\dfrac{\gamma_{n+1}}{\gamma_n \rho_n}\right) \dfrac{\langle W\delta^{(n)}, W\delta^{(n)}\rangle}{\langle W\delta^{(n-1)}, W\delta^{(n-1)}\rangle}\right]^{-1}, & n \geq 1, \end{cases}$$

and $\rho_1 = 1$. These formulæ involve a *symmetrization* matrix W that depends on the basic iterative method. (We will not go into the details of selecting W here but refer the reader to [2], for additional details.)

Summary. The primary computation for both of these two polynomial acceleration procedures are matrix-vector multiplications involving the iteration matrix G and a linear combinations of three vectors:

$$y \leftarrow Gx, \qquad y \leftarrow ax + bv + cz,$$

which are both vectorizable/parallelizable operations.

5. Vectorization. ITPACKV 2D [10] is an adaptation for vector supercomputers of the ITPACK 2C [4] package. When modifying the scalar package to create a vector version, only minor changes were made in the iterative algorithms. The primary changes in the vector version involved

- using a vectorizable data structure for storage of the coefficient matrix, and
- using a *wavefront* ordering to allow the SOR, SSORCG, and SSORSI routines to vectorize when the natural ordering was used.

In the vector version of the package, the data structure was chosen to coincide with the one used in the ELLPACK [15] software package. This storage format can not only efficiently store matrices arising from the discretization of partial differential equations on general domains as well as being flexible enough to accommodate matrix systems with various structures but it allows for greater efficiency of the iterative methods on vector computers. Also in the vector version of the package, the SOR and SSOR modules with the natural ordering permute the linear system according to wavefronts prior to applying these iterative procedures and then unpermutes it afterwards. The wavefront ordering is determined by finding sets of unknows that can be updated simultaneously. For example the unknowns along the diagonal of the grid would be the wavefronts for the five-point star operators on a rectangular mesh. For wavefronts, the vector lengths are dependent upon the sparsity pattern of the matrix as well as its size and are typically much smaller than the order of the system. The wavefront technique is used for improved vectorization and as such has been shown to be effective on vector supercomputers with small start-up times for their vector operations.

Matrix-Vector Multiplication. As we have shown, matrix-vector multiplication is one of the major components of the basic iterative methods in ITPACKV. The vectorization of this operation depends strongly on the choice of the sparse matrix storage format being used. In ITPACK 2C, a rowwise format was used involving three linear arrays **A**, **JA**, **IA**. With this data structure, the vector lengths may be quite small for many sparse matrices since they are directly related to the number of nonzeros in each row of the coefficient matrix. Using this storage scheme, matrix-vector multiplication involves dot-products and gather operations. Dot-products vectorize on most vector computers but often have a large start-up cost.

As an example, consider the matrix-vector multiplication $y \leftarrow Ax$ of the form

$$\begin{bmatrix} y_1 \\ y_2 \\ y_3 \\ y_4 \end{bmatrix} \leftarrow \begin{bmatrix} a_{11} & 0 & 0 & a_{14} \\ 0 & a_{22} & 0 & 0 \\ a_{31} & 0 & a_{33} & 0 \\ 0 & a_{42} & 0 & a_{44} \end{bmatrix} \begin{bmatrix} x_1 \\ x_2 \\ x_3 \\ x_4 \end{bmatrix}.$$

In ITPACK 2C, this system would be stored in three linear arrays as follows

$$\mathtt{A} = \begin{bmatrix} a_{11} & a_{14} & a_{22} & a_{31} & a_{33} & a_{42} & a_{44} \end{bmatrix},$$

$$\mathtt{JA} = \begin{bmatrix} 1 & 4 & 2 & 1 & 3 & 2 & 4 \end{bmatrix},$$

$$\mathtt{IA} = \begin{bmatrix} 1 & 3 & 4 & 6 & 8 \end{bmatrix}.$$

With this data structure, the nonzero elements of the matrix are stored rowwise in consecutive locations in the linear array **A**. The corresponding column numbers of these nonzeros are stored in the linear array JA of the same length. To distinguish between the entries in these arrays, a third linear array IA is needed containing pointers into A and JA indicating the beginning locations of each new row. Thus, $\mathtt{IA}(i+1)-\mathtt{IA}(i)$ is the number of nonzeros in the i-th row.

The matrix-vector product $y \leftarrow Ax$ for the example above can be formed as the sum of dot-products using the rowwise information in this storage scheme. It would appear as

$$\begin{bmatrix} y_1 \\ y_2 \\ y_3 \\ y_4 \end{bmatrix} \leftarrow \begin{bmatrix} a_{11}x_1 + a_{14}x_4 \\ a_{22}x_2 \\ a_{31}x_1 + a_{33}x_3 \\ a_{42}x_2 + a_{44}x_4 \end{bmatrix}$$

and this corresponds to the following

$$\mathtt{Y}(i) \leftarrow \sum_{k=\mathtt{IA}(i)}^{\mathtt{IA}(i+1)-1} \mathtt{A}(k) * \mathtt{X}(\mathtt{JA}(k)).$$

As mentioned previously, this involves short vectors and many dot-products.

In ITPACKV 2D, this system would be stored in two 2-dimensional arrays as follows:

$$\mathtt{COEF} = \begin{bmatrix} a_{11} & a_{14} \\ a_{22} & 0 \\ a_{33} & a_{31} \\ a_{44} & a_{42} \end{bmatrix}, \qquad \mathtt{JCOEF} = \begin{bmatrix} 1 & 4 \\ 2 & 2 \\ 3 & 1 \\ 4 & 2 \end{bmatrix}.$$

The operations $y \leftarrow Ax$ would be computed as

$$\begin{bmatrix} y_1 \\ y_2 \\ y_3 \\ y_4 \end{bmatrix} \leftarrow \begin{bmatrix} a_{11} \\ a_{22} \\ a_{33} \\ a_{44} \end{bmatrix} * \begin{bmatrix} x_1 \\ x_2 \\ x_3 \\ x_4 \end{bmatrix} + \begin{bmatrix} a_{14} \\ 0 \\ a_{31} \\ a_{42} \end{bmatrix} * \begin{bmatrix} x_4 \\ x_2 \\ x_1 \\ x_2 \end{bmatrix}.$$

Notice that this involves columns from array COEF and the entries from JCOEF indicate the corresponding x_i components. These operations can be computed by updating the vector y repeatedly to produce Ax. In general, the pseudocode for computing $y \leftarrow Ax$ would be as follows:

```
Y = 0
for j = 1,2,...,maxnz
   for i = 1,2,...,n
      Y(i) = Y(i) + COEF(i,j) * X(JCOEF(i,j))
   end
end
```

Here **maxnz** is the maximum number of nonzeros per row over all rows. Gather operations are needed to accumulate the X components and then a vector update operation is performed. Matrix-vector multiplication using this sparse storage format has the advantage that vectors lengths are longer (equal to n, the order of the system). Also, fewer gather operations and vector updates are needed than before (only **maxnz** for each).

The array Y in the algorithm above acts as an accumulation vector and, therefore, it should reside in the vector cache for the duration of the j loop, if one is available. Since Y is read once and written **maxnz** times, memory traffic between the main memory and the vector registers can be reduced by unrolling the j loop.

Forward or Backward Solves. Forward or backward solutions of sparse lower or upper triangular systems are major components of the **SOR** and **SSOR** routines with the natural ordering. These computational kernels can be vectorized using the idea of *computational wavefronts*.

For example using a wavefront re-ordering, the structure of the permuted system $(I - \widetilde{L} - \widetilde{U})x = b$ would be of the form

$$\begin{bmatrix} I_1 & -\widetilde{U}_{1,2} & \cdots & -\widetilde{U}_{1,m} \\ -\widetilde{L}_{2,1} & I_2 & \cdots & -\widetilde{U}_{2,m} \\ \vdots & \vdots & \ddots & \vdots \\ -\widetilde{L}_{m,1} & -\widetilde{L}_{m,2} & \cdots & I_m \end{bmatrix} \begin{bmatrix} X_1 \\ X_2 \\ \vdots \\ X_m \end{bmatrix} = \begin{bmatrix} B_1 \\ B_2 \\ \vdots \\ B_m \end{bmatrix},$$

where m is the number of wavefronts. The solution of $(I - \widetilde{L})x = y$ is a forward substitution of submatrices given by

$$X_i = Y_i + \sum_{j=1}^{i-1} \widetilde{L}_{i,j} X_j, \qquad (i = 1, 2, \ldots, m).$$

Similarly, the solution of $(I - \widetilde{U})x = y$ is a backward substitution given by

$$X_i = Y_i + \sum_{j=i+1}^{m} \widetilde{U}_{i,j} X_j, \qquad (i = m, m-1, \ldots, 1).$$

Hence, forward/backward solves can be be computed as a series of matrix-vector multiplications. While this vectorizes, the vector lengths are only the size of the wavefronts, which may be short.

In the case of solving a partial differential equation on a rectangle using a five-point finite-difference stencil, the computations can be ordered along the *diagonals* of the mesh. Since all nodes along a mesh diagonal can be updated simultaneously, they decouple and depend only upon the nodes along the previous lower diagonal. For an $m \times m$ mesh, the average vector length is $\approx m/2$ for forward/backward solves. Similarly in 3-dimensions using the seven-point finite-difference stencil on an $m \times m \times m$ box, all nodes lying on a plane diagonal of the mesh can be updated simultaneously with average vector lengths of $\approx m^2/3$.

In **ITPACKV 2D**, the computational wavefront concept was generalized to handle irregular grids and general operators so that the **SOR** and the **SSOR** routines using the natural ordering would vectorize. Given an ordering of the unknowns and a matrix structure, the package determines an equivalent wavefront ordering such that groups of consecutively numbered unknowns can be updated simultaneously. Details on the algorithms used can be found in [14]. Once a wavefront reordering is determined, the linear system is permuted prior to starting the iterative process so that unknowns in the same wavefront are in consecutive memory locations. After completion of all iterations, the system is permuted back to its original ordering. (Additional details on the vectorization effort are presented in [13].

6. Parallelization. In the vectorization of the ITPACK 2C package, an effort was made to identify and to exploit the fine-grain parallelism inherent in the algorithms. The package was then rewritten using vectorizable loops and renamed ITPACKV 2D. In many cases, the vector lengths are longer in this version of the package and significant parallel work can be done in the vectorizable loops. Thus, the basic idea behind the initial parallelization of the ITPACKV 2D package was to achieve course-grain parallelism by assigning portions of each vectorizable loop to available processors.

The guiding goals in the parallelization effort were

- to minimize the number of synchronization points, and hence, to maximize the amount of computation between these synchronization points, and
- to load-balance the work among available processors.

To achieve the first goal, synchronization points were used only when necessary (e.g., whenever the result of a dot-product was needed by all processors or whenever scalar code required execution on a single processor). To achieve the second goal, approximately equal portions of each vectorized loop were assigned to all available processors.

As a first step, the package was modified for execution on the following shared-memory parallel computers:

- Alliant FX/8 (8 processors)
- Sequent Symmetry (14 processors)

Since these computers use different parallelization commands, it was necessary to produce different versions of the package. The Alliant version of the code required the fewest modifications since compiler directives could be used to automatically partition the loops. However, the Sequent version required that each vectorizable loop be explicitly partitioned.

In the following sections, examples of parallel code for each of these computers will be given. The examples concern the parallelization of the matrix-vector product operation, $y \leftarrow Ax$, where A is stored in the ELLPACK sparse data format. This operation was used to vectorize each of the iterative methods in ITPACKV 2D and was the most important computational kernel for parallelization.

Alliant Version. On the Alliant computer, compiler directives were used for the segmentation of loops for parallel execution. The compiler automatically vectorizes and parallelizes single loops so there is no need to explicitly partition the loops. When loops are nested, the compiler attempts to vectorize the *inner* loop and parallelize the *outer* loop. This feature presents difficulties for load-balancing if the trip count of the outer loop does not match the number of available processors. In this case, a compiler directive was used to prevent the parallelization of the outer loop.

For example, computing the matrix-vector product $y \leftarrow Ax$ illustrates this principle:

```
          do 10 i = 1,n
             y(i) = 0.0
   10     continue
 cvd$ noconcur
          do 20 j = 1,maxnz
             do 15 i = 1,n
                y(i) = y(i) + coef(i,j)*x(jcoef(i,j))
   15        continue
   20     continue
```

Here the line **cvd$ noconcur** is a compiler directive used to prevent parallelization of loop 20. The inner loop 15 is then both vectorized and parallelized. The number of processors to be used can be specified at run time using a flag on the execution command. Each processor computes the matrix-vector product for a set of rows determined automatically at run time by the number of processors assigned. For example, if eight processors are available, then processor one computes the matrix-vector product for rows $1, 9, 17, \ldots$, processor two for rows $2, 10, 18, \ldots$, etc.

Sequent Version. For the Sequent version of the package, loops were explicitly broken into segments for parallel execution. This was done by introducing an outer loop that defines the segments and assigns them to available processors. Synchronization points were minimized by inserting within a single outer loop several inner loops containing no dependencies or needing only local information from a single processor. The **c$ doacross** compiler directive was used to parallelize the outer loops. For this directive, *shared* variables and *local* variables in the loop must be specified. The shared variables are known by all processors whereas the local variables are known only to a specific processor. The Sequent multiprocessor preprocessor rewrites the Fortran code so that each **c$ doacross** loop is converted to a subroutine.

An example of the use of this directive to compute $y \leftarrow Ax$ is given below. Each processor computes y_i for a range of rows determined by the endpoints **ist** and **ied**.

```
c$ doacross share (ntask,n,maxnz,y,x,coef,jcoef), local (i,j,ist,ied)
      do 30 k = 1,ntask
         ist = ((k-1)*n)/ntask + 1
         ied = (k*n)/ntask
         do 10 i = ist,ied
            y(i) = 0.0
10       continue
         do 20 j = 1,maxnz
            do 15 i = ist,ied
               y(i) = y(i) + coef(i,j)*x(jcoef(i,j))
15          continue
20       continue
30    continue
```

In this code, no synchronization point is needed between loops 10 and 20 since values in the vector **y** that are updated in loop 15 are initialized in loop 10 by the same processor.

7. Experimental Results. The problems used in the numerical experiments were the following:

2-D Model Problem

$$\begin{cases} u_{xx} + 2u_{yy} = 0, & (x,y) \in \Omega_2 = (0,1) \times (0,1), \\ u = 1 + xy, & (x,y) \in \delta\Omega_2. \end{cases}$$

The standard 5-point central finite-difference stencil was used with mesh sizes of $h = \frac{1}{30}$ and $\frac{1}{60}$, resulting in problem sizes of $N = 29^2 = 841$ and $59^2 = 3481$, respectively.

3-D Model Problem

$$\begin{cases} u_{xx} + 2u_{yy} + u_{zz} = 0, & (x,y,z) \in \Omega_3 = (0,1)^3, \\ u = 1, & \text{if } x = 0 \text{ or } y = 0 \text{ or } z = 0, \\ u_x = yz(1+yz), & \text{if } x = 1, \\ u_y = xz(1+xz), & \text{if } y = 1, \\ u_z = xy(1+xy), & \text{if } z = 1. \end{cases}$$

The standard 7-point central finite-difference stencil was used with mesh sizes of $h = \frac{1}{10}$ and $\frac{1}{16}$, resulting in problem sizes of $N = 10^3 = 1000$ and $16^3 = 4096$, respectively.

The stopping criterion used was the ITPACKV stopping test given by:

$$\frac{1}{1 - M_E} \frac{\|\delta^{(n)}\|_2}{\|u^{(n)}\|_2} < \zeta,$$

where $\delta^{(n)} = Gx^{(n)} + k - x^{(n)}$ is the current pseudo-residual, $u^{(n)}$ is the current solution vector, and M_E is an estimate of the maximum eigenvalue of the iteration matrix G. For all experiments, $\zeta = 10^{-6}$ was used. The use of this stringent stopping criterion required that double precision (64-bit) versions of parallel ITPACKV be used for both the Alliant and Sequent computers. In all of the tables, the number of iterations (Its.) for convergence of the routines with the stopping test above are given with the timing results reported in seconds. Both the natural ordering and a red-black ordering of the equations and unknowns were used. Adaptive procedures were used for all parameters except for the SSORCG and SSORSI runs under red-black ordering, for which $\omega = 1$ was fixed. The numbers in parentheses are the speedups for the multiprocessor code with different numbers of processors (p) over the multiprocessor code run in uniprocessor mode ($p = 1$).

Alliant FX/8 Results. The following conclusions can be drawn from Tables 1, 2, 3, and 4:

- The JCG and JSI routines under natural ordering and all routines under red-black ordering parallelized well. The SOR, SSORCG, and SSORSI routines under natural ordering did not do as well. For these three routines, the length of the vector operations in the forward and backward SOR passes were short. Also, a synchronization point was needed for each diagonal of the mesh since the computations for each wavefront of nodes required the results from the previous wavefront.
- The SI routines often had better speedups than the corresponding CG routines because they had fewer dot-products, and hence, fewer synchronization points.
- Comparing Tables 1 and 2, it is apparent that increasing the problem size for the 2-D problem improved the speedup ratios. This is to be expected since the ratio of parallel work to scalar work increases as the problem size is increased. Comparing Tables 3 and 4, increasing the problem size for the 3-D problem had mixed results on the speedup ratios. The large problem had worse speedup ratios than the small problem for small numbers of processors and better speedup ratios for large numbers of processors.
- Comparing Tables 1 and 3, it is apparent that the speedup ratios are better for the 3-D problem than for the 2-D problem, even though the problem sizes

are approximately the same. This may be due to the fact that the parallelized matrix-vector product is a more significant part of an iteration for the 3-D problem than for the 2-D problem. It can also be seen that the speedup ratios for the **SOR**, **SSORCG**, and **SSORSI** routines under natural ordering are significantly better for the 3-D problem than for the 2-D problem. This may be due to the fact that there are fewer wavefronts (and hence fewer synchronization points) for the 3-D problem than for the 2-D problem.

- The fastest routine depends upon the number of processors used and the problem size. Consider Tables 1 and 3. For both the small 2-D and 3-D problems, under the natural ordering, the **SSORCG** and **SSORSI** routines are the fastest with one processor ($p = 1$), but the **JCG** routine is the fastest using eight processors ($p = 8$). For one processor, the superior convergence rates for the **SSORCG** and **SSORSI** methods makes them preferable to the **JCG** method. For eight processors, however, the superior parallelization of the **JCG** method outweighs its inferior convergence rate to make it preferable to the **SSORCG** and **SSORSI** methods. Now consider Tables 2 and 4. For the larger 2-D problem, the **SSORSI** method is the preferable method under natural ordering for all the processors used due to its far superior convergence rate compared to the **JCG** method. For the larger 3-D problem, the results are mixed since the difference in convergence rates is not so great between the **JCG** and **SSOR** methods.
  Under red-black ordering, the **RSCG** method is the preferable method for each of the four problems for any number of processors.

Sequent Symmetry Results. For the Sequent Symmetry, only the small 2-D and 3-D problems were considered due to the Sequent's slower processor. The experiments were run in single-user mode using up to 13 processors. The following conclusions can be drawn from Tables 5 and 6:

- An apparently super-linear speedup was obtained for some runs using 2 and 3 processors. This may be due to inflated times for uniprocessor execution ($p = 1$) because of a greater number of cache misses.
- As with the Alliant runs, the speedup ratios were greatest for the **JCG** and **JSI** routines under natural ordering and all routines under red-black ordering. The speedup ratios were small for the **SOR**, **SSORCG**, and **SSORSI** methods under natural ordering.
- Comparing Table 1 with Table 5 and Table 3 with Table 6, it can be seen that the Alliant was faster than the Sequent for the same problem but had lower speedup ratios for the same number of processors. The individual processors on the Alliant are vector processors while those on the Sequent are scalar processors. With the slower processors, the penalty for doing synchronizations on the Sequent are much smaller than for the Alliant.
- Comparing Table 5 with Table 6, it can be seen that the speedups are greater for the 3-D problem than for the 2-D problem. This is especially true for the **SOR**, **SSORCG**, and **SSORSI** methods under natural ordering.

8. General Conclusions. The following conclusions concerning parallel **ITPACKV** can be drawn from the runs on the two parallel computers used in this experiment:

- **ITPACKV 2D** was effectively parallelized for certain shared-memory parallel computers with relatively small numbers of processors. However, the **SOR**, **SSORCG**, and SSORSI routines under natural ordering did not parallelize as well due to the short vector lengths and large number of synchronization points in the SOR forward and backward passes.

- Since the SI routines compute fewer dot-products, they can become competitive with the corresponding CG routines as the number of processors grows.
- The preferred routine depends upon the number of available processors and the problem size. The SSORCG and SSORSI routines may be preferred for low numbers of processors since they often perform fewer iterations than the remaining routines in ITPACKV. However, the superior parallelization of the JCG, JSI, RSCG, and RSSI routines may make them the preferred routines for moderate numbers of processors and small problem sizes.
- In its present form, ITPACKV 2D is not likely to be the most efficient implementation of iterative algorithms for solving large sparse linear systems on parallel computers since the package was developed for sequential computers and then it was parallelized. Since there is substantial scalar (uniprocessor) work in these algorithms for estimating parameters, new or modified algorithms need to be developed with parallelization in mind from the beginning.

Acknowledgements. The authors wish to acknowledge the Center for Numerical Analysis and the Computation Center at The University of Texas at Austin (UT Austin) for their support. Also, we would like to mention the organizations at UT Austin that were helpful and the computer systems that were used in the development and testing of the parallel ITPACKV software package as well as in the writing of this paper: the Texas Institute for Computational Mechanics (TICOM) for the use of the Alliant FX/8 (Concentrix 5.502, FORTRAN V4.2.40) and the Computer Sciences Department for the use of the Sequent Symmetry (DYNIX V3.0.12, FORTRAN 77 V3.2) plus the use of SUN 3 workstations. Suggestions for improving this paper were received from Irina Mukherjee and Malathi Ramdas. We would also like to acknowledge a most helpful referee's report resulting in a revised and improved paper.

This work was supported, in part, by the National Science Foundation under grant CCR-8518722, the Department of Energy under grant DE-FG05-87ER25048, and Cray Research, Inc., under grant LTR DTD with The University of Texas at Austin.

REFERENCES

[1] Grimes, R. G., D. R. Kincaid, and D. M. Young. "ITPACK 2.0 user's guide." Report CNA-150, Center for Numerical Analysis, University of Texas at Austin, August, 1979.

[2] Hageman, L. A., and D. M. Young. *Applied Iterative Methods*, New York, Academic Press, 1981.

[3] Kincaid, D. R., and D. M. Young. "Survey of iterative methods." *Encyclopedia of Computer Sciences and Technology*, Vol. 13, (J. Belzer, A. G. Holzman, and A. Kent, eds.), M. Dekker, New York, 1979, pp. 354–391.

[4] Kincaid, D. R., J. R. Respess, D. M. Young, and R. G. Grimes. "ITPACK 2C: A FORTRAN package for solving large sparse linear systems by adaptive accelerated iterative methods." *ACM Transactions on Mathematical Software*, Vol. 8, No. 3, September 1982, pp. 302–322.

[5] Kincaid, D. R., and T. C. Oppe. "ITPACK on supercomputers." *Numerical Methods* (eds. V. Pereyra and A. Reinoza), pp. 151–161. New York: Springer-Verlag, Lecture Notes in Mathematics, No. 1005, 1983. Also Report CNA-178, Center for Numerical Analysis, University of Texas at Austin, September 1982.

[6] Kincaid, D. R., T. C. Oppe, and D. M. Young. "Adapting ITPACK routines for use on a vector computer." *Proceedings of the Symposium on Cyber 205 Applications*, The International

Symposium on Vector Processing Applications, Colorado State University, Fort Collins, Colorado, August 12–13, 1982. Also Report CNA-177, Center for Numerical Analysis, University of Texas at Austin, August 1982.

[7] Kincaid, D. R., T. C. Oppe, and D. M. Young. "ITPACKV 2C user's guide." Report CNA-191, Center for Numerical Analysis, University of Texas at Austin, February 1984.

[8] Kincaid, D. R., T. C. Oppe, and D. M. Young. "Vector computations for sparse linear systems." *SIAM Journal for Algebraic and Discrete Methods*, Vol. 7, No. 1, January 1986, pp. 99–112.

[9] Kincaid, D. R., and D. M. Young. "A brief review of the ITPACK project." *Journal of Computational and Applied Mathematics*, Vol. 24, 1988, pp. 33–54.

[10] Kincaid, D. R., T. C. Oppe, and D. M. Young. "ITPACKV 2D user's guide." Report CNA-232, Center for Numerical Analysis, University of Texas at Austin, May, 1989.

[11] Kincaid, D. R., and T. C. Oppe. "Chapter 9: Some parallel algorithms on the four processor Cray X-MP4 supercomputer," *Parallel Supercomputing: Methods, Algorithms, and Applications*, (ed. G. F. Carey), Chichester, England: Wiley & Sons, 1989, pp. 121–134.

[12] Oppe, T. C., and D. R. Kincaid. "The performance of ITPACK on vector computers for solving large sparse linear systems arising in sample oil reservoir simulation problems." *Communications in Applied Numerical Methods*, Vol. 3, No. 1, 1987, pp. 23–29. Also Report CNA-200 (Revised), Center for Numerical Analysis, University of Texas at Austin, April 1986.

[13] Oppe, T. C. "The vectorization of ITPACK 2C." *International Journal for Numerical Methods in Engineering*, Vol. 27, 1989, pp. 571–588.

[14] Oppe, T. C. *The Iterative Solution of Large Sparse Linear Systems Using Vector Computers.* Ph. D. dissertation, University of Texas at Austin, May 1990.

[15] Rice, J. R., and R. F. Boisvert. *Solving Elliptic Problems Using ELLPACK.* New York: Springer-Verlag, 1985.

[16] Young, D. M. *Iterative Solution of Large Linear Systems.* New York: Academic Press, 1971.

[17] Young, D. M., T. C. Oppe, D. R. Kincaid, and L. J. Hayes. "On the use of vector computers for solving large sparse linear systems." *Proceedings of the Conference on Vector and Parallel Processors for Scientific Computation*, held at Academia dei Lincei Palazzina dell' Auditoria, via della Lungara 230, Rome, Italy, May 27–29, 1985. Also Report CNA-199, Center for Numerical Analysis, University of Texas at Austin, May 1985.

TABLE 1
Alliant FX/8, 2-D Problem, $h = \frac{1}{30}$

Routine	Its.	Number of Tasks							
		p=1	p=2	p=3	p=4	p=5	p=6	p=7	p=8
Natural Ordering									
JCG	101	0.904	0.520	0.439	0.329	0.295	0.276	0.264	0.231
		(1.000)	(1.738)	(2.059)	(2.748)	(3.064)	(3.275)	(3.424)	(3.913)
JSI	180	1.340	0.727	0.561	0.419	0.357	0.337	0.299	0.270
		(1.000)	(1.843)	(2.389)	(3.198)	(3.754)	(3.976)	(4.482)	(4.963)
SOR	96	1.056	0.727	0.628	0.579	0.587	0.575	0.563	0.553
		(1.000)	(1.453)	(1.682)	(1.824)	(1.799)	(1.837)	(1.876)	(1.910)
SSORCG	23	0.713	0.495	0.439	0.386	0.374	0.363	0.358	0.346
		(1.000)	(1.440)	(1.624)	(1.847)	(1.906)	(1.964)	(1.992)	(2.061)
SSORSI	28	0.735	0.518	0.459	0.393	0.386	0.378	0.370	0.360
		(1.000)	(1.419)	(1.601)	(1.870)	(1.904)	(1.944)	(1.986)	(2.042)
Red-Black Ordering									
JCG	101	0.960	0.562	0.477	0.362	0.329	0.309	0.297	0.263
		(1.000)	(1.708)	(2.013)	(2.652)	(2.918)	(3.107)	(3.232)	(3.650)
JSI	180	1.397	0.770	0.599	0.455	0.392	0.372	0.333	0.308
		(1.000)	(1.814)	(2.332)	(3.070)	(3.564)	(3.755)	(4.195)	(4.536)
SOR	103	0.865	0.476	0.373	0.288	0.251	0.239	0.215	0.197
		(1.000)	(1.817)	(2.319)	(3.003)	(3.446)	(3.619)	(4.023)	(4.391)
SSORCG	50	0.978	0.555	0.451	0.349	0.307	0.282	0.267	0.238
		(1.000)	(1.762)	(2.169)	(2.802)	(3.186)	(3.468)	(3.663)	(4.109)
SSORSI	89	1.441	0.811	0.651	0.460	0.399	0.376	0.340	0.309
		(1.000)	(1.777)	(2.214)	(3.133)	(3.612)	(3.832)	(4.238)	(4.663)
RSCG	51	0.466	0.283	0.231	0.195	0.177	0.170	0.159	0.149
		(1.000)	(1.647)	(2.017)	(2.390)	(2.633)	(2.741)	(2.931)	(3.128)
RSSI	88	0.666	0.371	0.299	0.240	0.209	0.202	0.181	0.174
		(1.000)	(1.795)	(2.227)	(2.775)	(3.187)	(3.297)	(3.680)	(3.828)

TABLE 2
Alliant FX/8, 2-D Problem, $h = \frac{1}{60}$

Routine	Its.	Number of Tasks							
		p=1	p=2	p=3	p=4	p=5	p=6	p=7	p=8
Natural Ordering									
JCG	205	7.474	4.200	2.992	2.321	2.032	1.765	1.759	1.522
		(1.000)	(1.780)	(2.498)	(3.220)	(3.678)	(4.235)	(4.249)	(4.911)
JSI	376	12.288	6.607	4.858	3.691	3.093	2.816	2.676	2.431
		(1.000)	(1.860)	(2.529)	(3.329)	(3.973)	(4.364)	(4.592)	(5.055)
SOR	205	8.464	5.351	4.080	3.460	3.198	3.098	2.961	2.884
		(1.000)	(1.582)	(2.075)	(2.446)	(2.647)	(2.732)	(2.858)	(2.935)
SSORCG	36	4.077	2.698	2.058	1.794	1.653	1.581	1.528	1.441
		(1.000)	(1.511)	(1.981)	(2.273)	(2.466)	(2.579)	(2.668)	(2.829)
SSORSI	39	3.901	2.532	1.970	1.746	1.590	1.527	1.479	1.426
		(1.000)	(1.541)	(1.980)	(2.234)	(2.453)	(2.555)	(2.638)	(2.736)
Red-Black Ordering									
JCG	205	7.714	4.350	3.121	2.451	2.159	1.892	1.882	1.651
		(1.000)	(1.773)	(2.472)	(3.147)	(3.573)	(4.077)	(4.099)	(4.672)
JSI	376	12.588	6.726	4.976	3.808	3.219	2.947	2.802	2.553
		(1.000)	(1.872)	(2.530)	(3.306)	(3.911)	(4.271)	(4.483)	(4.931)
SOR	201	7.263	4.078	2.889	2.188	1.889	1.725	1.621	1.503
		(1.000)	(1.781)	(2.514)	(3.319)	(3.845)	(4.210)	(4.481)	(4.832)
SSORCG	101	7.949	4.740	3.231	2.631	2.260	2.040	1.902	1.706
		(1.000)	(1.677)	(2.460)	(3.021)	(3.517)	(3.897)	(4.179)	(4.659)
SSORSI	179	12.410	7.162	4.897	4.077	3.362	3.026	2.818	2.607
		(1.000)	(1.733)	(2.534)	(3.044)	(3.691)	(4.101)	(4.404)	(4.760)
RSCG	103	3.449	1.923	1.436	1.128	1.028	0.905	0.872	0.813
		(1.000)	(1.794)	(2.402)	(3.058)	(3.355)	(3.811)	(3.955)	(4.242)
RSSI	187	5.584	3.145	2.215	1.775	1.526	1.401	1.251	1.227
		(1.000)	(1.776)	(2.521)	(3.146)	(3.659)	(3.986)	(4.464)	(4.551)

TABLE 3
Alliant FX/8, 3-D Problem, $h = \frac{1}{10}$

Routine	Its.	Number of Tasks							
		p=1	p=2	p=3	p=4	p=5	p=6	p=7	p=8
Natural Ordering									
JCG	52	0.731	0.395	0.298	0.243	0.217	0.199	0.183	0.163
		(1.000)	(1.851)	(2.453)	(3.008)	(3.369)	(3.673)	(3.995)	(4.485)
JSI	125	1.503	0.785	0.550	0.444	0.399	0.354	0.323	0.277
		(1.000)	(1.915)	(2.733)	(3.385)	(3.767)	(4.246)	(4.653)	(5.426)
SOR	71	1.176	0.727	0.574	0.512	0.483	0.459	0.443	0.418
		(1.000)	(1.618)	(2.049)	(2.297)	(2.435)	(2.562)	(2.655)	(2.813)
SSORCG	15	0.727	0.453	0.382	0.334	0.322	0.298	0.287	0.275
		(1.000)	(1.605)	(1.903)	(2.177)	(2.258)	(2.440)	(2.533)	(2.644)
SSORSI	33	1.211	0.729	0.585	0.533	0.486	0.455	0.437	0.413
		(1.000)	(1.661)	(2.070)	(2.272)	(2.492)	(2.662)	(2.771)	(2.932)
Red-Black Ordering									
JCG	52	0.821	0.459	0.356	0.296	0.270	0.251	0.234	0.214
		(1.000)	(1.789)	(2.306)	(2.774)	(3.041)	(3.271)	(3.509)	(3.836)
JSI	125	1.595	0.849	0.608	0.498	0.453	0.406	0.376	0.327
		(1.000)	(1.879)	(2.623)	(3.203)	(3.521)	(3.929)	(4.242)	(4.878)
SOR	70	0.974	0.528	0.386	0.318	0.294	0.262	0.247	0.219
		(1.000)	(1.845)	(2.523)	(3.063)	(3.313)	(3.718)	(3.943)	(4.447)
SSORCG	26	0.884	0.476	0.383	0.309	0.287	0.248	0.233	0.209
		(1.000)	(1.857)	(2.308)	(2.861)	(3.080)	(3.565)	(3.794)	(4.230)
SSORSI	64	1.723	0.891	0.660	0.571	0.472	0.415	0.386	0.337
		(1.000)	(1.934)	(2.611)	(3.018)	(3.650)	(4.152)	(4.464)	(5.113)
RSCG	27	0.465	0.272	0.214	0.185	0.172	0.160	0.155	0.143
		(1.000)	(1.710)	(2.173)	(2.514)	(2.703)	(2.906)	(3.000)	(3.252)
RSSI	65	0.844	0.466	0.345	0.289	0.266	0.237	0.227	0.204
		(1.000)	(1.811)	(2.446)	(2.920)	(3.173)	(3.561)	(3.718)	(4.137)

TABLE 4
Alliant FX/8, 3-D Problem, $h = \frac{1}{16}$

Routine	Its.	Number of Tasks							
		p=1	p=2	p=3	p=4	p=5	p=6	p=7	p=8
Natural Ordering									
JCG	85	5.004	2.796	2.043	1.666	1.425	1.273	1.144	1.012
		(1.000)	(1.790)	(2.449)	(3.004)	(3.512)	(3.931)	(4.374)	(4.945)
JSI	193	10.320	5.652	4.207	3.218	2.812	2.459	2.194	1.931
		(1.000)	(1.826)	(2.453)	(3.207)	(3.670)	(4.197)	(4.704)	(5.344)
SOR	106	6.854	4.261	3.076	2.642	2.417	2.210	2.091	1.907
		(1.000)	(1.609)	(2.228)	(2.594)	(2.836)	(3.114)	(3.278)	(3.594)
SSORCG	21	3.743	2.308	1.826	1.561	1.439	1.338	1.262	1.190
		(1.000)	(1.622)	(2.050)	(2.398)	(2.601)	(2.797)	(2.966)	(3.145)
SSORSI	37	5.700	3.481	2.628	2.299	2.057	1.915	1.768	1.667
		(1.000)	(1.637)	(2.169)	(2.479)	(2.771)	(2.977)	(3.224)	(3.419)
Red-Black Ordering									
JCG	85	5.497	3.134	2.351	1.941	1.693	1.526	1.398	1.263
		(1.000)	(1.754)	(2.338)	(2.832)	(3.247)	(3.602)	(3.932)	(4.352)
JSI	193	10.783	5.939	4.497	3.484	3.067	2.710	2.434	2.181
		(1.000)	(1.816)	(2.398)	(3.095)	(3.516)	(3.979)	(4.430)	(4.944)
SOR	109	6.539	3.765	2.700	2.197	1.897	1.706	1.592	1.409
		(1.000)	(1.737)	(2.422)	(2.976)	(3.447)	(3.833)	(4.107)	(4.641)
SSORCG	43	5.987	3.378	2.510	2.018	1.798	1.625	1.482	1.333
		(1.000)	(1.772)	(2.385)	(2.967)	(3.330)	(3.684)	(4.040)	(4.491)
SSORSI	95	11.058	6.288	4.475	3.683	3.191	2.935	2.551	2.339
		(1.000)	(1.759)	(2.471)	(3.002)	(3.465)	(3.768)	(4.335)	(4.728)
RSCG	43	2.790	1.659	1.258	1.051	0.937	0.869	0.803	0.753
		(1.000)	(1.682)	(2.218)	(2.655)	(2.978)	(3.211)	(3.474)	(3.705)
RSSI	94	5.043	2.994	2.156	1.760	1.531	1.362	1.266	1.152
		(1.000)	(1.684)	(2.339)	(2.865)	(3.294)	(3.703)	(3.983)	(4.378)

TABLE 5

Sequent Symmetry, 2-D Problem, $h = \frac{1}{30}$

Routine	Its.	Number of Tasks												
		p=1	p=2	p=3	p=4	p=5	p=6	p=7	p=8	p=9	p=10	p=11	p=12	p=13
Natural Ordering														
JCG	101	7.75 (1.00)	3.85 (2.01)	2.66 (2.91)	2.06 (3.76)	1.72 (4.51)	1.48 (5.24)	1.32 (5.87)	1.19 (6.51)	1.10 (7.05)	1.03 (7.52)	0.97 (7.99)	0.92 (8.42)	0.87 (8.91)
JSI	180	10.73 (1.00)	5.34 (2.01)	3.58 (3.00)	2.73 (3.93)	2.24 (4.79)	1.89 (5.68)	1.66 (6.46)	1.48 (7.25)	1.35 (7.95)	1.25 (8.58)	1.15 (9.33)	1.08 (9.94)	1.01 (10.62)
SOR	96	6.78 (1.00)	4.12 (1.65)	3.18 (2.13)	2.72 (2.49)	2.47 (2.74)	2.29 (2.96)	2.18 (3.11)	2.09 (3.24)	2.05 (3.31)	1.98 (3.42)	1.95 (3.48)	1.90 (3.51)	1.89 (3.59)
SSORCG	23	5.03 (1.00)	2.96 (1.70)	2.24 (2.25)	1.89 (2.66)	1.68 (2.99)	1.54 (3.27)	1.45 (3.47)	1.38 (3.64)	1.34 (3.75)	1.30 (3.87)	1.26 (3.99)	1.24 (4.06)	1.21 (4.16)
SSORSI	28	5.16 (1.00)	3.03 (1.70)	2.32 (2.22)	1.96 (2.63)	1.76 (2.93)	1.62 (3.19)	1.53 (3.37)	1.47 (3.51)	1.42 (3.63)	1.37 (3.77)	1.34 (3.85)	1.32 (3.91)	1.29 (4.00)
Red-Black Ordering														
JCG	101	7.86 (1.00)	4.17 (1.88)	2.88 (2.73)	2.23 (3.52)	1.86 (4.23)	1.60 (4.91)	1.43 (5.50)	1.30 (6.05)	1.19 (6.61)	1.12 (7.02)	1.05 (7.49)	1.00 (7.86)	0.95 (8.27)
JSI	180	11.01 (1.00)	5.79 (1.90)	3.91 (2.82)	2.99 (3.68)	2.43 (4.53)	2.06 (5.34)	1.81 (6.08)	1.61 (6.84)	1.47 (7.49)	1.35 (8.16)	1.25 (8.81)	1.17 (9.41)	1.10 (10.01)
SOR	103	6.27 (1.00)	3.43 (1.83)	2.42 (2.59)	1.90 (3.30)	1.55 (4.05)	1.34 (4.68)	1.18 (5.31)	1.06 (5.92)	0.97 (6.46)	0.90 (6.97)	0.84 (7.46)	0.79 (7.94)	0.75 (8.36)
SSORCG	50	8.18 (1.00)	4.27 (1.92)	2.98 (2.74)	2.31 (3.54)	1.90 (4.31)	1.64 (4.99)	1.45 (5.64)	1.32 (6.20)	1.21 (6.76)	1.13 (7.24)	1.06 (7.72)	1.00 (8.18)	0.95 (8.61)
SSORSI	89	11.96 (1.00)	6.19 (1.93)	4.29 (2.79)	3.31 (3.61)	2.71 (4.41)	2.32 (5.16)	2.03 (5.89)	1.84 (6.50)	1.66 (7.20)	1.54 (7.77)	1.43 (8.36)	1.34 (8.93)	1.27 (9.42)
RSCG	51	3.38 (1.00)	1.80 (1.88)	1.27 (2.66)	1.01 (3.35)	0.86 (3.93)	0.75 (4.51)	0.68 (4.97)	0.62 (5.45)	0.58 (5.83)	0.55 (6.15)	0.52 (6.50)	0.51 (6.63)	0.48 (7.04)
RSSI	88	4.85 (1.00)	2.52 (1.92)	1.72 (2.82)	1.34 (3.62)	1.10 (4.41)	0.96 (5.05)	0.85 (5.71)	0.77 (6.30)	0.71 (6.83)	0.65 (7.46)	0.61 (7.95)	0.58 (8.36)	0.55 (8.82)

TABLE 6

Sequent Symmetry, 3-D Problem, $h = \frac{1}{10}$

Routine	Its.	Number of Tasks												
		p=1	p=2	p=3	p=4	p=5	p=6	p=7	p=8	p=9	p=10	p=11	p=12	p=13
Natural Ordering														
JCG	52	6.23 (1.00)	3.04 (2.05)	2.06 (3.02)	1.60 (3.89)	1.31 (4.76)	1.13 (5.51)	0.99 (6.29)	0.89 (7.00)	0.81 (7.69)	0.74 (8.42)	0.69 (9.03)	0.65 (9.58)	0.62 (10.05)
JSI	125	11.99 (1.00)	5.90 (2.03)	3.97 (3.02)	3.01 (3.98)	2.46 (4.87)	2.08 (5.76)	1.81 (6.62)	1.60 (7.49)	1.44 (8.33)	1.31 (9.15)	1.21 (9.91)	1.12 (10.71)	1.05 (11.42)
SOR	71	7.64 (1.00)	4.22 (1.81)	3.05 (2.50)	2.49 (3.07)	2.14 (3.57)	1.91 (4.00)	1.75 (4.37)	1.64 (4.66)	1.53 (4.99)	1.46 (5.23)	1.41 (5.42)	1.36 (5.62)	1.32 (5.79)
SSORCG	15	4.82 (1.00)	2.67 (1.81)	1.93 (2.50)	1.57 (3.07)	1.35 (3.57)	1.21 (3.98)	1.11 (4.34)	1.03 (4.68)	0.97 (4.97)	0.92 (5.24)	0.88 (5.48)	0.86 (5.60)	0.83 (5.81)
SSORSI	33	8.35 (1.00)	4.55 (1.84)	3.27 (2.55)	2.65 (3.15)	2.27 (3.68)	2.02 (4.13)	1.84 (4.54)	1.72 (4.85)	1.61 (5.19)	1.52 (5.49)	1.46 (5.72)	1.40 (5.96)	1.37 (6.09)
Red-Black Ordering														
JCG	52	6.54 (1.00)	3.44 (1.90)	2.35 (2.78)	1.82 (3.59)	1.50 (4.36)	1.29 (5.07)	1.15 (5.69)	1.03 (6.35)	0.95 (6.88)	0.88 (7.43)	0.83 (7.88)	0.77 (8.49)	0.73 (8.96)
JSI	125	12.47 (1.00)	6.43 (1.94)	4.34 (2.87)	3.29 (3.79)	2.68 (4.65)	2.25 (5.54)	1.98 (6.30)	1.75 (7.13)	1.59 (7.84)	1.45 (8.60)	1.34 (9.31)	1.25 (9.98)	1.17 (10.66)
SOR	70	7.17 (1.00)	3.79 (1.89)	2.65 (2.71)	2.04 (3.51)	1.69 (4.24)	1.44 (4.98)	1.28 (5.60)	1.15 (6.23)	1.04 (6.89)	0.95 (7.55)	0.89 (8.06)	0.84 (8.54)	0.79 (9.08)
SSORCG	26	6.92 (1.00)	3.63 (1.91)	2.54 (2.72)	1.97 (3.51)	1.61 (4.30)	1.40 (4.94)	1.23 (5.63)	1.11 (6.23)	1.01 (6.85)	0.93 (7.44)	0.88 (7.86)	0.83 (8.34)	0.79 (8.76)
SSORSI	64	13.50 (1.00)	6.98 (1.93)	4.82 (2.80)	3.69 (3.66)	3.01 (4.49)	2.56 (5.27)	2.25 (6.00)	2.01 (6.72)	1.81 (7.46)	1.66 (8.13)	1.54 (8.77)	1.44 (9.38)	1.36 (9.93)
RSCG	27	3.30 (1.00)	1.75 (1.89)	1.23 (2.68)	0.98 (3.37)	0.82 (4.02)	0.71 (4.65)	0.64 (5.16)	0.59 (5.59)	0.54 (6.11)	0.50 (6.60)	0.48 (6.88)	0.46 (7.17)	0.43 (7.67)
RSSI	65	6.16 (1.00)	3.16 (1.95)	2.18 (2.83)	1.69 (3.64)	1.40 (4.40)	1.20 (5.13)	1.07 (5.76)	0.96 (6.42)	0.87 (7.08)	0.80 (7.70)	0.75 (8.21)	0.71 (8.68)	0.67 (9.19)

On the sparsity patterns of hierarchical finite element matrices

J. Maubach*

Faculty of Mathematics and Computer Science,
University of Nijmegen
Toernooiveld 5, 6525 ED Nijmegen
The Netherlands

Abstract

The performance of preconditioned conjugate gradient methods for the solution of a linear system of equations $H\underline{x} = \underline{b}$ depends strongly on the quality of the preconditioner. In many applications the system of equations to be solved originates from a partial differential equation discretized by the choice of an initial mesh geometry, a meshrefinement technique and a type of finite element basisfunctions. In general the resulting matrix is a sparse matrix which sparsity pattern only depends on the discretization choices. For the construction of a preconditioner only the matrix entries are needed, but investigations so far clearly have shown that taking into account additionally the discretization choices via the sparsity pattern leads to more effective preconditioning techniques, of which many exist.

Relatively simple techniques like ILU and SSOR take the matrix entries into account but disregard the sparsity pattern. Somewhat more effective techniques like Block-Incomplete ones make use of the sparsity patterns regularity resulting from a regular mesh geometry. In addition multi-grid techniques take into account that a fine mesh geometry is obtained by the refinement of a coarse mesh. Intermediate 'levels' of refinement are distinguished and used explicitly. More flexible are algebraic multilevel preconditioners which assign a 'level' to each individual degree of freedom.

As the sparsity pattern is of importance for the construction of good preconditioners it is analysed for the hierarchical matrix H resulting from a given discretization. The hierarchy is induced by the mesh refinement method applied. It is shown that the sparsity pattern is irregular but well structured in general and a simple refinement method is presented which enables a compact storage and quick retrieval of the matrix entries in the computers memory. An upperbound for the C.-B.-S. scalar for this method is determined to demonstrate that it is well suited for multilevel preconditioning and it is shown to have satisfying angle bounds. Further it turns out that the hierarchical matrix may be partially constructed in parallel, is block structured and shows block decay rates.

Keywords: Sparsity structure, Adaptive mesh refinement, Hierarchical finite elements, Error indication, Parallel computing

AMS(MOS) subject classifications: 65N30, 65N50, 65W05, 65F10, 65F10, 65F50

---

* research supported by the Netherlands Organization for Scientific Research N.W.O.

1 Introduction

At present the use of hierarchical finite element basisfunctions, in combination with adaptive mesh refinement for the solution of elliptic boundary value problems, has been investigated thoroughly (see e.g. [9], [10], [12], [17], [21], [22], [23] and [28]) to show that this approach is eminently applicable to problems with complicated boundary geometry. Most approaches investigated sofar implicitly use the hierarchical matrix H, resulting from a mesh obtained by an adaptive mesh refinement method, via the relationship $H = I^tAI$ as in [28] where A is the standard nodal matrix on the mesh mentioned and I is the mesh geometry dependent lower triangular identity transformation from the hierarchical to the standard nodal basis. As multiplications with I^{-t} and I are easy to perform, solving $H\underline{x} = \underline{b}$ for a given vector $\underline{b}$ with the use of the equivalent system $A(I\underline{x}) = I^{-t}\underline{b}$ as in [18] has several advantages. It is possible to do the assembly of, storage of and multiplication with A elementwise (see e.g. [14]) and the number of matrix entries to be stored will be proportional to the number of unknowns. On the other hand I is often defined recursively prohibiting a parallel multiplication (one can use domain decomposition techniques to overcome this) and the use of elementwise stored data induces much indirect computer addressing each time a multiplication is performed, slowing down the overall computational performance.

All approaches mentioned avoid an analysis of *the sparsity pattern* J, the set of couplings (i,j) corresponding to possible non-zero entries h_{ij} of the hierarchical matrix H, which is more complicated in the hierarchical than in the ordinary finite element case. Because one does not know the sparsity patterns structure a straightforward row-wise ordered storage technique (see e.g. [2]) is used; for each degree of freedom i, the numbers j for which $(i,j) \in J$ are stored in an integer row. This technique is often quite expensive due to a generally large number of shifts needed to store the numbers j in an increasing order. To check whether $(i,j) \in J$ for given i and j, e.g. occuring during a pointwise incomplete factorization of H, is costly because this involves searching through an integer row.

This paper will show that for the discretization choices of a given mesh refinement technique, a coarse initial mesh and type of finite element basisfunctions it is possible to determine the sparsity patterns structure of the resulting hierarchical matrix by studying the changes in J caused by refinement of a single triangle. This shows for frequently used refinement techniques that the sparsity pattern has a (sometimes even binary) tree structure which can be stored in a row-wise ordered manner without any shifting and such that checking whether $(i,j) \in J$ for given j involves at most one '`if...then`' instruction, independent on the dimension of the matrix and the number of refinements applied to the initial coarse mesh.

Unfortunately the number of possible non-zero entries of H is bounded above by $O(kN)$, k the number of refinements and N the number of unknowns, whence a matrix vector multiplication will not be of optimal computational complexity $O(N)$ if the number of refinement levels is unrestricted. However, this is partially compensated for by the fact that in general the entries h_{ij} are of order $O(\sqrt{2}^{-|(p-q)|})$ if the related basisfunctions are of level p respectively q as is shown in [20]. Also, the storage of the hierarchical matrix in the computer memory allows for the direct application of many well-known preconditioning techniques.

Since the sparsity patterns structure is depending on the type of local refinement technique used, some types of refinement techniques will be studied in more detail. A refinement technique introduced to this end is the *newest vertex bisection* technique proposed by Sewell [25], [26] and adapted by Mitchell [21], similar to the *longest edge bisection* method of Rivara [23] and [24]. The difference between the methods is that for a given triangle the former method bisects the edge opposite a specific vertex whereas the latter method bisects its longest edge. Unlike the *regular mesh refinement* used by [4], Bank [10], [15], [17] and Yserentant [28] these bisection methods do not create children congruent to their parents.

The main advantage of the technique proposed is its simplicity. Mitchells technique is a one fase recursive technique, which keeps all triangles *compatible* (at most one neighbour along each of the three edges) at all times contrary to Rivara, who has to enforce compatibility after each refinement by refining an additional number of triangles, and Bank and Yserentant [9], who even need a third fase in which the bisections used to enforce compatibility are removed. Further advantages are that there is no need to compute side lengths as in Rivaras refinement algorithm and that the number of different angles created during repeatedly refinement at most is eight times the number of triangles in the initial coarse mesh whence a properly chosen initial mesh can automatically avoid bad angles. The newest vertex bisection method is generalizable to domains in more than two space dimensions and for higher order finite element basisfunctions.

As the sparsity pattern is analysed bearing the construction of an effective multilevel preconditioner in mind, angle bounds for the refinement methods above are investigated and the Cauchy-Buniakowskii-Schwarz scalar γ^2 is derived. The results obtained demonstrate that mesh refinement methods leading to a well structured sparsity pattern in addition can be well suited for multilevel preconditioning.

The remainder of the paper is organized as follows. The newest vertex mesh refinement method, which will be used to demonstrate the results of the sparsity pattern analysis in the view of its simplicity, is introduced in section 2. In section 3 a general sparsity pattern analysis is presented for various mesh refinement techniques. The number of the hierarchical matrix entries as well as the storage of the complete hierarchical matrix in the computer memory are investigated briefly in section 4. Thereafter section 5 considers angle bounds for the mesh refinement methods studied and provides estimates for γ^2 in order to show that also the newest vertex refinement method, having one of the simplest structured sparsity patterns possible, may be used for the construction of multilevel preconditioners. Finally in section 6 sparsity patterns resulting from some discretizations of example partial differential equations are presented and in section 7 some conclusions are drawn.

2 The newest vertex mesh refinement

The regular as well as the bisection mesh refinement method have several basic properties in common which are essential with respect to the sparsity pattern analysis. This section introduces the newest vertex mesh refinement and considers these properties.

A basic building block of each mesh refinement algorithm is the method to divide a triangle, which is supposed to be a topologically open subset of the computational domain Ω. In the newest vertex bisection case a triangle Δ of a certain *level* $l(\Delta)$ has three vertices labeled 1, 2 and 3 of which vertex 1, denoted by $\hat{\varphi}_\Delta$, is called the *newest vertex*. By definition the edge opposite this vertex is called the *base* and the unique triangle sharing this base is said to be the *neighbour*, if this exists (see fig. 1.1). The bisection of a triangle Δ always takes place from its newest vertex to the midpoint of its base. The vertex created will be the newest vertex of both *children* Δ_i $(i = 1, 2)$, which are defined properly, i.e.,

$$\Delta = \cup_i \Delta_i, \quad \Delta_i \cap \Delta_j = \emptyset \text{ if } i \neq j. \tag{2.1}$$

By definition the children have the unique *parent* $P(\Delta_i) = \Delta$ and are of level

$$l(\Delta_i) = l(\hat{\varphi}_{\Delta_i}) = l(\Delta) + 1\,, \tag{2.2}$$

as are the vertex and line created. All vertices, lines and triangles of the initial coarse triangulation have level 0. For future use consider the subsequent *ancestors*

$$P^0(\Delta) = \Delta \text{ and } P^k(\Delta) = \underbrace{(P \circ \cdots \circ P)}_{k>0}(\Delta) \quad \forall_{k=1,\ldots,l(\Delta)}$$

of a triangle $\Delta \in T$.

As mentioned before, one of the difficulties in mesh refinement is that of maintaining the compatibility of the triangulation. Therefore consider the refinement of a triangle in a given mesh in which all triangles are *compatibly divisible*. This means that for every triangle either its base is lying at the boundary or its neighbours base coincides with its own base. Now the refinement of a triangle is defined by the following recursive algorithm:

- If it is compatibly divisible then bisect it and its neighbour, if this latter exists.
- If it is not compatibly divisible then first refine its neighbour and thereafter refine itself.

Fig. 2.1 shows for a coarse initial mesh where all triangles are compatibly divisible some refinement steps, including the refinement of a not compatibly divisible triangle, and figs. 2.5 and 2.9 show an already refined mesh. It is of course of importance that the refinement algorithm above will always yield a compatible mesh and that the recursion involved will be of finite length and not too large, as is shown by the following lemma.

Lemma 2.1

(i) Given any initial triangulation, there exists a choice of newest vertices such that every triangle is compatibly divisible.

(ii) The length of the recursion involved with the refinement of an arbitrary triangle is bounded by its level plus 1.

(iii) For each pair of compatibly divisible triangles Δ_1 and Δ_2, $l(\Delta_1) = l(\Delta_2)$.

Proof. See Mitchell [22] for a proof in full detail.

•

As the sparsity pattern analysis also depends on the type of finite element basisfunctions used consider a two dimensional open and bounded polygonal domain Ω with an initial coarse compatibly divisible triangulation and standard nodal piecewise linear or quadratic finite element basisfunctions, called *nodes* for abbreviation. The *hierarchical* finite element basis is constructed step by step as follows.

Each refinement of a triangle as above creates some children and a vertex $\hat{\varphi}$. At this new vertex the new hierarchical node φ is defined as is usual in finite element methods, piecewise linear or quadratic, and only locally non-zero on the surrounding triangles (see e.g. [2], [13], [16], [27] and [29]). In more detail, for future use, let the *support* and the *base* of a node φ be defined by respectively

$$D_\varphi := \{x \in \Omega : \varphi(x) \neq 0\}^*$$
$$B_\varphi := \{\Delta : D_\varphi \cap \Delta \neq \emptyset \wedge l(\Delta) = l(\varphi)\}$$

where for a set X, X^* denotes the open part of its topological closure. Further, as is standard in finite elements, the base is supposed to be such that there exists a scalar $c \in \mathbb{N}$, independent on φ, such that $|B_\varphi| \leq c$. Further, the refinement is supposed to lead to a compatible mesh, which is formulated by

$$\Delta \in B_\varphi \Rightarrow \hat{\varphi} \in v(\Delta) \wedge l(\Delta) = l(\varphi), \tag{2.3}$$

where $v(\Delta)$ denotes the set of vertices of Δ. This implies in addition that if $\Delta \in B_\varphi$, then $\hat{\varphi}$ is one its vertices. Note that according to the definition of the base B_φ, all elements (triangles) are of the same level $l(\varphi)$, whence their parents must also be of the same level. For the newest vertex bisection method this is shown by the previous lemma.

The created vertex $\hat{\varphi}$ and corresponding node φ are assigned a unique number i, i.e., $\varphi = \varphi_i$, such that after the local refinement for all nodes

$$l(\varphi_i) \leq l(\varphi_j) \Leftrightarrow i \leq j, \tag{2.4}$$

where $l(\varphi) = l(\hat{\varphi})$ by definition. This leads to a naturally block structured hierarchical matrix, where the blocks are determined by the levels of the refinement. It is advised to number the vertices and corresponding nodes of the initial coarse grid such that the resulting coarse mesh matrix is as sparse as possible (see e.g. [2] for an algorith to this end). Fig. 1.6 shows a one dimensional hierarchical basis and fig. 2.2 shows the numbers of the nodes for the mesh given in fig. 2.1 (see also figs. 2.5 and 2.9).

Define the *triangle set* τ_k to be the set of all triangles of level k, $k \geq 0$. A refined triangle in τ_k is not removed from this set and can not be refined again, its children belong to τ_{k+1}. Further, *node set* v_k resp. *vertex set* $\underline{v}_k$ will contain all nodes resp. vertices of level k, $k \geq 0$. Let $V \equiv V_k := \cup_{i=0}^{k} v_k$, $T \equiv T_k := \cup_{i=0}^{k} \tau_k$, $n_k := |v_k|$ and let $N \equiv N_k := |V_k|$ be the total number of nodes in V. Define $\underline{V} \equiv \underline{V}_k = \cup_{i=0}^{k} \underline{v}_k$ and let $W \equiv W_k$ denote the span of the nodes in V_k, which is identical for the hierarchical and the standard nodal basis. Note that the subset of T of triangles without children has exactly $\frac{1}{2}(T + T_0)$ elements in the case of bisection refinement. This subset suffices for the definition of a standard nodal basis.

3 A sparsity pattern analysis

After defining the sparsity pattern with the use of some of the definitions from section 2, some elementary results valid for mesh refinement methods satisfying the conditions (2.1) up to (2.4) are proven. Thereafter these results are used to examine the sparsity pattern structure for a general case. Afterwards the general result obtained is considered for the regular and newest vertex bisection refinement in more detail.

Now consider the definition of the sparsity pattern resulting from a chosen discretization. A node φ_j is said to be *coupled* with node φ_i and $(i,j) \in N \times N$ is said to be a *coupling* iff $D_{\varphi_i} \cap D_{\varphi_j} \neq \emptyset$. The *sparsity pattern* is defined as the set of couplings

$$J = \{(i,j) \in N \times N : \varphi_i, \varphi_j \in V \wedge D_{\varphi_j} \cap D_{\varphi_i} \neq \emptyset\},$$

depending on the mesh geometry and the finite element basisfunctions used but *not* on the partial differential equation discretized. Note that the sparsity pattern is a symmetric subset of $N \times N$ by definition, whence it suffices to examine the *couplings sets* $C(\varphi_i)$

$$C(\varphi_i) := \{\varphi_j \in V : D_{\varphi_j} \cap D_{\varphi_i} \neq \emptyset \wedge j \leq i\} \quad \forall_{i=1,\dots,N}.$$

containing all nodes coupled to and of lower level than φ_i. Before proceeding to the main analysis consider the following lemma providing some basic and simple tools.

Lemma 3.1 *A mesh refinement method satisfying (2.1) – (2.4) leads to a mesh T with*

- *The intersection of every $\Delta_1, \Delta_2 \in T$ is either a subset of one of them, or empty*

$$\forall_{\Delta_1,\Delta_2 \in T}[\Delta_1 \subset \Delta_2 \vee \Delta_2 \subset \Delta_1 \vee \Delta_1 \cap \Delta_2 = \emptyset]. \tag{3.1}$$

- *If $\Delta_1, \Delta_2 \in T$ and $\Delta_1 \cap \Delta_2 \neq \emptyset$ then*

$$\begin{cases} l(\Delta_1) < l(\Delta_2) \Rightarrow \Delta_1 \supset \Delta_2 \\ l(\Delta_1) = l(\Delta_2) \Rightarrow \Delta_1 = \Delta_2 \\ l(\Delta_1) > l(\Delta_2) \Rightarrow \Delta_1 \subset \Delta_2 \end{cases}. \tag{3.2}$$

- *All triangles $\Delta, \tilde{\Delta} \in T$ satisfy*

$$\begin{cases} l(P^{l(\Delta)-i}(\Delta)) = i \quad \forall_{i=0,1,\dots,l(\Delta)} \\ \Delta \subset \tilde{\Delta} \Leftrightarrow P^{l(\Delta)-l(\tilde{\Delta})}(\Delta) = \tilde{\Delta} \end{cases}. \tag{3.3}$$

Proof. The results above follow directly with the use of combinations of the equations (2.1) up to (2.4).

•

Intuitively, considering the definition of the nodes in the previous section, a couplings set of a given node φ will contain nodes defined on ancestors of the triangles belonging the base of φ. For locally refined grids, consisting of a few triangles, this can easily be verified by hand, but in the general case one must exploit some of the basic properties provided in the previous section in order to prove this.

Now let $p(\varphi)$ denote the number of parents which created a node φ, note that

$$p(\varphi) = |\{P(\Delta) : \Delta \in B_\varphi\}|,$$

and define the sets D and E for all triangles $\Delta \in T$ by

$$\begin{aligned} D(\Delta) &:= \{\mu \in V : D_\mu \cap \Delta \neq \emptyset \wedge l(\mu) \leq l(\Delta)\} \\ E(\Delta) &:= \{\mu \in V : D_\mu \cap \Delta \neq \emptyset \wedge l(\mu) = l(\Delta)\}. \end{aligned} \tag{3.4}$$

Because the couplings set of a level zero node is easy to determine only couplings sets of created nodes are considered in the following theorem.

Theorem 3.1 *For all nodes $\varphi \in V - v_0$ and all $\tilde{\Delta} \in T$ containing D_φ, $C(\varphi)$ is the union of three disjunct sets*

$$C(\varphi) = \bigcup_{\Delta \in B_\varphi} E(\Delta) \cup \bigcup_{\Delta \in B_\varphi} \bigcup_{k=1}^{l(\Delta)-l(\tilde{\Delta})-1} E(P^k(\Delta)) \cup \bigcup_{k=0}^{l(\tilde{\Delta})} E(P^k(\tilde{\Delta})). \tag{3.5}$$

Proof. Let $\varphi \in V - v_0$. Nodes created after φ have a number greater than the number of φ according to (2.4), whence no such nodes are added to φ-s couplings set after its creation. The creation of new triangles also does not influence the old nodes and couplings whence $C(\varphi)$ can be examined under the assumption that φ is the last node created.

Note that, due to (2.4), directly after a node $\varphi_i \equiv \varphi$ has been created, this node has the highest possible node number i, leading to

$$\begin{aligned} C(\varphi_i) &= \{\varphi_j \in V : D_{\varphi_j} \cap D_{\varphi_i} \neq \emptyset \wedge j \leq i\} \\ &= \{\varphi_j \in V : D_{\varphi_j} \cap D_{\varphi_i} \neq \emptyset\} \\ &= \{\mu \in V : D_\mu \cap D_\varphi \neq \emptyset\} \\ &= \bigcup_{\Delta \in B_\varphi} \{\mu \in V : D_\mu \cap \Delta \neq \emptyset\} \end{aligned}$$

because $D_\varphi = (\cup\{\Delta : \Delta \in B_\varphi\})^*$. Furthermore, suppose $D_\mu \cap \Delta \neq \emptyset$, then there exists $\tilde{\Delta} \in B_\mu$ such that $\tilde{\Delta} \cap \Delta \neq \emptyset$. Property (2.1) now implies $\Delta \subset \tilde{\Delta}$ whence according to (3.2) $l(\mu) = l(\tilde{\Delta}) \leq l(\Delta)$. This leads to

$$C(\varphi) = \bigcup_{\Delta \in B_\varphi} \{\mu \in V : D_\mu \cap \Delta \neq \emptyset \wedge l(\mu) \leq l(\Delta)\} = \bigcup_{\Delta \in B_\varphi} D(\Delta).$$

Now consider the sets $D(\Delta)$. Note first of all that

$$\forall_{\Delta \in T - \tau_0} \forall_{\mu \in V} [D_\mu \cap \Delta \neq \emptyset \wedge l(\mu) \leq l(P(\Delta)) \Leftrightarrow D_\mu \cap P(\Delta) \neq \emptyset \wedge l(\mu) \leq l(P(\Delta))] \tag{3.6}$$

because for any $\Delta \in T - \tau_0$ and node $\mu \in V$

$$\begin{aligned} & D_\mu \cap P(\Delta) \neq \emptyset \wedge l(\mu) \leq l(P(\Delta)) \Leftrightarrow \\ & \exists_{\tilde{\Delta} \in B_\mu} [\tilde{\Delta} \cap P(\Delta) \neq \emptyset \wedge l(\tilde{\Delta}) \leq l(P(\Delta))] \Leftrightarrow \\ & \exists_{\tilde{\Delta} \in B_\mu} [P(\Delta) \subset \tilde{\Delta} \wedge l(\tilde{\Delta}) \leq l(P(\Delta))] \Rightarrow \\ & \exists_{\tilde{\Delta} \in B_\mu} [\Delta \subset \tilde{\Delta} \wedge l(\tilde{\Delta}) \leq l(P(\Delta))] \Rightarrow \\ & \exists_{\tilde{\Delta} \in B_\mu} [\tilde{\Delta} \cap \Delta \neq \emptyset \wedge l(\tilde{\Delta}) \leq l(P(\Delta))] \Leftrightarrow \\ & D_\mu \cap \Delta \neq \emptyset \wedge l(\mu) \leq l(P(\Delta)) \end{aligned}$$

due to (3.2), (2.3) and $\Delta \subset P(\Delta)$. The opposite direction follows with the use of $\Delta \cap \tilde{\Delta} \neq \emptyset \Rightarrow P(\Delta) \cap \tilde{\Delta} \neq \emptyset$. Therefore the relationship between the sets $D(\Delta)$ and $E(\Delta)$ is, according to (3.6), given by

$$\begin{aligned} D(\Delta) &= \{\mu \in V : D_\mu \cap \Delta \neq \emptyset \wedge l(\mu) \leq l(\Delta)\} \\ &= \{\mu \in V : D_\mu \cap \Delta \neq \emptyset \wedge l(\mu) < l(\Delta)\} \cup \\ &\quad \{\mu \in V : D_\mu \cap \Delta \neq \emptyset \wedge l(\mu) = l(\Delta)\} \\ &= \{\mu \in V : D_\mu \cap \Delta \neq \emptyset \wedge l(\mu) \leq l(P(\Delta))\} \cup E(\Delta) \\ &= \{\mu \in V : D_\mu \cap P(\Delta) \neq \emptyset \wedge l(\mu) \leq l(P(\Delta))\} \cup E(\Delta) \\ &= D(P(\Delta)) \cup E(\Delta) \end{aligned} \tag{3.7}$$

whence according to $D(P^{l(\Delta)}(\Delta)) = E(P^{l(\Delta)}(\Delta))$ and relation (3.7)

$$D(\Delta) = \bigcup_{k=0}^{l(\Delta)} E(P^k(\Delta)) \text{ and } C(\varphi) = \bigcup_{\Delta \in B_\varphi} \bigcup_{k=0}^{l(\Delta)} E(P^k(\Delta)).$$

Note that the sets $E(P^k(\Delta)), k = 0, 1, \ldots, l(\Delta)$ are mutually disjunct because they contain nodes of different levels.

Now suppose that there exists a triangle $\tilde{\Delta} \in T$ such that $D_\varphi \subset \tilde{\Delta}$ then due to (3.3)

$$\forall_{\Delta \in B_\varphi} [P^{l(\Delta)-l(\tilde{\Delta})}(\Delta) = \tilde{\Delta}] \Rightarrow \forall_{\Delta \in B_\varphi} \forall_{k=0,l(\tilde{\Delta})} [P^{l(\Delta)-l(\tilde{\Delta})+k}(\Delta) = P^k(\tilde{\Delta})]$$

whence

$$\begin{aligned} C(\varphi) &= \bigcup_{\Delta \in B_\varphi} D(\Delta) = \bigcup_{\Delta \in B_\varphi} \bigcup_{k=0}^{l(\Delta)} E(P^k(\Delta)) = \bigcup_{k=0}^{l(\Delta)} \bigcup_{\Delta \in B_\varphi} E(P^k(\Delta)) \\ &= \bigcup_{\Delta \in B_\varphi} E(\Delta) \cup \bigcup_{\Delta \in B_\varphi} \bigcup_{k=1}^{l(\Delta)-l(\tilde{\Delta})-1} E(P^k(\Delta)) \cup \bigcup_{k=0}^{l(\tilde{\Delta})} E(P^k(\tilde{\Delta})). \end{aligned}$$

Clearly three different parts can be distinguished. The first part of the union, called *head*, consists of a union of sets $E(\Delta)$ with a non-empty intersection containing the nodes created simultaneously with φ. The second part, called *middle* part, consists of sets $E(\Delta)$ which may overlap partially but the last part, called *tail* part, consists of mutually disjunct sets $E(\Delta)$.

Now consider the existence of such a triangle $\tilde{\Delta} \in T$. If $p(\varphi) = 1$ then there exists a unique parent $\tilde{\Delta}$ of level $l(\Delta) - 1$ which contains D_φ whence the above formula reduces to

$$C(\varphi) = \bigcup_{\Delta \in B_\varphi} E(\Delta) \cup \bigcup_{k=1}^{l(\tilde{\Delta})} E(P^k(\tilde{\Delta})).$$

If $p(\varphi) = 2$ and if φ is created at a line of level $m > 0$, then clearly

$$\exists_{\tilde{\Delta} \in T_{m-1}} [D_\varphi \subset \tilde{\Delta}]$$

whence

$$C(\varphi) = \bigcup_{\Delta \in B_\varphi} E(\Delta) \cup \bigcup_{\Delta \in B_\varphi} \bigcup_{k=1}^{l(\Delta)-l(\tilde{\Delta})-1} E(P^k(\Delta)) \cup \bigcup_{k=0}^{l(\tilde{\Delta})} E(P^k(\tilde{\Delta})).$$

For nodes with two parents, which are created at a line of level 0 the couplings set has only a head and middle part, the tail part has length zero. Note that this derivation is not restricted to a specific discretization.

•

The theorem above shows that indeed all nodes coupled to a certain given node φ are defined on one of the ancestors of a triangle belonging to the base of φ. In addition, it shows that it is sufficient to take into account only those nodes defined on a vertex of an ancestor $P^k(\Delta)$ which are an element of $E(P^k(\Delta))$ for given k and triangle $\Delta \in B_\varphi$. Obviously the sparsity pattern is structured with respect to the level structure of the mesh refinement under consideration.

In order to be able to estimate the number of couplings $|J|$, i.e., the number of matrix entries, consider the sets $E(\Delta)$, for certain $\Delta \in T$, in more detail.

Lemma 3.2 *For all $\Delta \in T$ the set $E(\Delta)$ satisfies*

$$E(\Delta) = \{\mu \in v(\Delta) : l(\mu) = l(\Delta)\}$$

whence there exist positive scalars a and b such that for all created nodes $\varphi \in V - v_0$

$$|C(\varphi)| \leq al(\varphi) + b$$

the length of the couplings set is bounded linearly by the level.

Proof. Consider the first part and suppose $\mu \in E(\Delta)$. Then $D_\mu \cap \Delta \neq \emptyset$ so there exists a $\tilde{\Delta} \in B_\mu$ such that $\tilde{\Delta} \cap \Delta \neq \emptyset$. Further $l(\tilde{\Delta}) = l(\mu) = l(\Delta)$, according to (3.2) resulting in $\Delta = \tilde{\Delta} \in B_\mu$ whence due to (2.3) $\mu \in v(\Delta)$. The opposite direction is trivial. According to the characterization of the E sets above $|E(\Delta)| \leq 3$ for all triangles $\Delta \in T$. The couplings sets for nodes $\varphi \in V - v_0$ satisfy, according to (3.7) and (2.3),

$$\begin{aligned} |C(\varphi)| &= \left| \bigcup_{\Delta \in B_\varphi} \bigcup_{k=0}^{l(\Delta)} E(P^k(\Delta)) \right| \leq \left| \bigcup_{\Delta \in B_\varphi} E(\Delta) \right| + 3p(\varphi)(l(\varphi) + 1) \\ &= 3c + 3p(\varphi)(l(\varphi) + 1) \end{aligned}$$

whence $a = 3p(\varphi)$ and $b = 3(c + p(\varphi))$ may be taken.

•

This lemma provides an upper bound for the number of entries of the hierarchical matrix for a general refinement technique satisfying the conditions in section 2. In order to obtain sharper estimates for specific refinement techniques, consider again the sets $E(\Delta)$, first for the regular mesh refinement technique considered by Bank and Yserentant in [28]. Their technique distincts three different types of triangle refinement (see fig. 1.7) called *red* of the first and second kind and *green*, necessary to maintain the compatibility of the mesh. Here $|E(\Delta)|$ depends on the triangle, for the red refinement $E(\Delta) = 3$ for the triangle Δ created in the middle of its parent and $E(\Delta) = 2$ for the triangles Δ created along the boundaries (see e.g. fig. 1.9). In the green bisection case $E(\Delta) = 1$ for both children as will be demonstrated below.

For the bisection refinement of Mitchell and Rivara, (the support of the created nodes differs from the regular refinement case as is shown in figs. 1.8 and 1.9), $|E(\Delta)| = 1$ for all $\Delta \in T - \tau_0$, as is a direct consequence of the lemma below.

Lemma 3.3 *The newest vertex mesh refinement implies*

(i) A triangle Δ of level $l(\Delta) \geq 1$ has exactly

one vertex of level $l(\Delta)$, its newest vertex $\hat{\varphi}_\Delta$

one vertex of level $l(\Delta) - 1$, the vertex $\hat{\varphi}_{P(\Delta)}$ and

one vertex of level less than $l(\Delta) - 1$ if $l(\Delta) > 1$ or of level 0 if $l(\Delta) = 1$.

(ii) A compatibly divisible pair of triangles Δ_1 and Δ_2 of level $l > 1$ have exactly

two vertices of level l, both newest vertices $\hat{\varphi}_{\Delta_1}$ and $\hat{\varphi}_{\Delta_2}$

one common vertex of level $l - 1$, $\hat{\varphi}_{P(\Delta_1)} = \hat{\varphi}_{P(\Delta_2)}$

one common vertex of level less than $l-1$

(iii) *Each created vertex $\hat{\varphi}_\Delta$ is situated on at most 7 different lines of which one of level less than $l(\Delta)$, the base at which $\hat{\varphi}_\Delta$ is created two of level $l(\Delta)$, created together with $\hat{\varphi}_\Delta$ and at most four of level $l(\Delta)+1$, created afterwards eventually.*

(iv) *If a vertex $\hat{\varphi}_\Delta$ is created at a line l_m of level $m>0$ then $\hat{\varphi}_{P^{2k}(\Delta)} \in l_m$ for all $0 \le 2k \le l(\hat{\varphi}_\Delta)-m$ where $l(\hat{\varphi}_\Delta)-m \ge 0$ is always even.*

Proof. The first statement is obviously true for $l(\Delta)$ equal to 1 or 2. Suppose it is true for triangles Δ of a certain level $l(\Delta)>1$ and take a triangle Δ with vertices $\hat{\varphi}_\Delta, \hat{\varphi}_{P(\Delta)}$ and $\hat{\mu}$ with $l(\hat{\mu}) < l(\Delta)-1$. The refinement of this triangle leads to the creation of one vertex $\hat{\varphi}$ of level $l(\Delta)+1$ and two children Δ_1 and Δ_2 such that $v(\Delta_1) = \{\hat{\varphi}, \hat{\varphi}_\Delta, \hat{\varphi}_{P(\Delta)}\}$ resp. $v(\Delta_2) = \{\hat{\varphi}, \hat{\varphi}_\Delta, \mu\}$. Clearly the vertices of both children satisfy the conditions posed on their level, whence by induction the statement is proved.

Note that two compatibly divisible triangles Δ_1 and Δ_2 of level k have four vertices always including the newest vertices $\hat{\varphi}_{\Delta_1}$ and $\hat{\varphi}_{\Delta_2}$ of level k and the $\hat{\varphi}_{P(\Delta_1)}$ and $\hat{\varphi}_{P(\Delta_2)}$ of level $k-1$. If $\hat{\varphi}_{P(\Delta_1)} \neq \hat{\varphi}_{P(\Delta_2)}$ then the refinement led to two triangles with one vertex of level k and two vertices of level $k-1$ which is impossible for $k>1$. For $k=1$ the statement does not hold, see fig. 1.5 for a counter example.

The third statement follows with induction, analogous to the first, so consider the last statement and suppose that $\hat{\varphi}_{P^{2k}(\Delta)}$ is created at a line l_m of level $l(\hat{\varphi}_{P^{2k}(\Delta)}) > l(l_m) = m > 0$. According to $l(\hat{\varphi}_{P^{2k}(\Delta)}) = l(\Delta)-2k > m \ge 1$, i.e., $0 \le 2k < l(\Delta)-m$ and according to (i) (the level of l_m is positive) also vertex $\hat{\varphi}_{P^{2k+2}(\Delta)}$ is situated at l_m. Furthermore, the row $\hat{\varphi}_\Delta, \hat{\varphi}_{P\Delta}, \hat{\varphi}_{P^1(\Delta)}, \ldots,$ must contain one of the endpoints of l_m, which is of level m. Therefore there exists a non-negative scalar k such that $l(\hat{\varphi}_\Delta)-m = 2k$, whence $l(\hat{\varphi}_\Delta)-m$ is even.

•

Theorem 3.1 in combination with lemma 3.3(ii) show that, for the newest vertex mesh refinement technique, the basisfunctions belonging to τ_k are not mutually coupled, i.e., $C(\varphi) \cap \tau_{l(\varphi)} = \{\varphi\}$ for all $k>0$. This means that the resulting block structured hierarchical matrix (see (2.4)) will have diagonal blocks being diagonal matrices. Since this refinement technique leads to a matrix with a simply structured sparsity pattern it is even possible to give a precize upperbound for the number of couplings $|C(\varphi)|$ of a node φ. To this end define the function $1_{\{m=0\}}$ on $\mathbb{N}$ to obtain value 1 if $m=0$ and 0 otherwise and consider

Theorem 3.2 *If $\varphi \in V - v_0$ is created at a line of level m then*

$$\begin{cases} |C(\varphi)| = l(\varphi)+3 & \text{if } p(\varphi)=1 \\ |C(\varphi)| = \frac{3}{2}l(\varphi)+3-\frac{1}{2}m+(1-\frac{1}{2}\mathrm{mod}(l(\varphi),2))1_{\{m=0\}} & \text{if } p(\varphi)=2. \end{cases} \tag{3.8}$$

Proof. Let $\varphi \in V - v_0$ have one parent whence automatically $m=0$. Because φ is created there exists a triangle $\Delta \in T-\tau_0$ such that $\varphi \equiv \varphi_\Delta$. Defining the *path* of this newest vertex φ_Δ by the row of nodes

$$\varphi_\Delta, \varphi_{P(\Delta)}, \varphi_{P^2(\Delta)}, \ldots, E(P^{l(\Delta)}(\Delta)) \tag{3.9}$$

then according to theorem 3.1 and lemma 3.2 the couplings set $C(\varphi)$ will exactly contain all elements of this row. All nodes in the couplings set are of different level, $|E(\Delta)|=1$ for all $\Delta \in T-\tau_0$ according to the previous lemma and the nodes on the coarsest mesh are defined standard nodally leading to the desired result.

If φ has two parents then according to theorem 3.1 $C(\varphi)$ contains exactly φ itself, at most $2=p(\varphi)$ ancestors at each level $0<k<l(\varphi)$ and at most 4 nodes of level 0. Lemma 3.2(iv) shows that for a node with two different parents $\Delta_1, \Delta_2 \in V$ the middle part of the couplings set will have several overlapping sets $E(P^{2k}(\Delta_i))$. The elementary counting necessary to obtain the second result above is left to the interested reader. Note that for this type of refinement $|C(\varphi)| \le \frac{3}{2}l(\varphi)+4$ for all nodes $\varphi \in V - v_0$.

•

As an example some graphs of paths of nodes of the meshes in figs. 2.2 and 2.5 are provided in figs. 2.3 resp. 2.6. Consider fig. 2.8 with initial coarse mesh $\tau_0 = \{\{\hat{\varphi}_1, \hat{\varphi}_2, \hat{\varphi}_4\}, \{\hat{\varphi}_3, \hat{\varphi}_4, \hat{\varphi}_2\}\}$ and fig. 2.3 as examples for which theorem 3.2 can easily be verified. Note that the first elements of the path of a vertex are not always the vertex itself and the vertices of its parent(s). This is shown for the path belonging to node φ_{19} in fig. 2.6, where parent vertex $\hat{\varphi}_7$ is visited after non-parent vertices $\hat{\varphi}_{10}$ and $\hat{\varphi}_{11}$. However, the theorem assures that by following the path all coupled nodes, also those defined at the vertices of the parent(s) will be visited eventually. The bisection refinement around a singularity is shown in fig. 1.5 where $C(\varphi_i) = i$ for all nodes φ_i situated on the line $y = x$ with $i > 4$. The standard nodal basis sparsity pattern is a subset of that of the hierarchical case due to the definition of the basisfunctions in both cases.

In this section it has been demonstrated that the sparsity patterns structure is only depending on the triangle ancestor hierarchy induced by the refinement strategy chosen. For certain types of refinement, like the newest vertex refinement technique, it has been shown that one can determine all coupled nodes by simply following paths of ancestor triangles.

4 Storage of the hierarchical matrix

As is shown the sparsity pattern of a matrix is a very irregular but structured subset of $N \times N$ in general. First it will be shown that the couplings set J can be stored efficiently due to this structure and thereafter the number of the matrix entries $|J|$ is computed for two examples.

The ordinary row-wise ordered storage scheme can be described with the use of a map $\tilde{M}: J \to \mathbb{N}$ as follows. For each $(i,j) \in J$ the number j is stored in a row of integers at position $\tilde{M}(i,j) = g(i) + f(j)$ where f and g are defined by

$$\begin{cases} g(1) = 0 \\ g(i) = g(i-1) + |C(\varphi_{i-1})|,\ i \geq 2 \end{cases} \quad \text{and } f(j) \in \{1, \ldots, |C(\varphi_i)|\}$$

for all $i,j \in \{1, \ldots, N\}$. Consider in the following the newest vertex bisection refinement, without loss of generality. In this case, according to theorem 3.1 and 3.2, a node φ_i maximally is coupled to itself, two nodes of level m for all $0 < m < l(\varphi_i)$ and four nodes of level 0. A special row-wise ordered storage scheme can be defined with the use of the map $M: J \to \mathbb{N}$, $M(i,j) = g(i) + f(j) + 3$ where f and g are defined by

$$\begin{cases} g(1) = 1 \\ g(i) = g(i-1) + 2l(\varphi_{i-1}) + 4,\ i \geq 2 \end{cases}$$

$$\begin{array}{llll} f(j) & \in \{-3, -2, -1, 0\} & \text{if} & 0 = l(\varphi_j) \\ & \in \{2l(\varphi_j) - 1, 2l(\varphi_j)\} & \text{if} & 0 < l(\varphi_j) \\ & = \{2l(\varphi_i) - 1\} & \text{if } l(\varphi_j) = & l(\varphi_i) \end{array}$$

for all $i,j \in \{1, \ldots, N\}$. As an example consider figs. 2.4 and 2.7 where for the meshes in fig. 2.2 resp. 2.5. for each node φ_i seperately the corresponding piece of the row is shown in a table, taking $g(i) = 1$ for all i for simplicity. The length of the couplings sets of nodes of level zero is typically 4 for a uniform coarse triangulation of the unit-square (see e.g. [2] or fig. 3.2 where the vertices are numbered from bottom to top and left to right) whence the proposed storage scheme can also be used for these nodes as $4 \leq 2 \cdot 0 + 4$.

The advantage of the special scheme is that for created nodes φ_i, φ_j checking whether $(i,j) \in J$ only involves at most one `'if...then'` instruction contrary to at most ${}^2\log|C(\varphi_i)|$ of such instructions in the ordinary case. The disadvantage is that $|M(J)| > |\tilde{M}(J)| = |J|$ but one can consider the computing speed gained more important then the loss of some computer memory.

The number of possible non-zero entries of the matrix resulting from the disretization is equal to

$$|J| \equiv |J_k| = 2 \cdot \sum_{i=1}^{N_k} |C(\varphi_i)| - N_k =: 2E_k - N_k \tag{4.1}$$

according to the definition of C. Therefore J is determined by E_k.

Theorem 4.1 *If the initial coarse mesh is as shown in the first picture of fig. 3.1 then*

- *after k succesive uniform newest vertex bisection refinement steps as shown in fig. 3.1*

$$N_k = \begin{cases} (2^{\frac{k}{2}} + 1)^2 & \text{if } k \text{ even} \\ (2^{\frac{k-1}{2}} + 1)^2 + 2^{k-1} & \text{elsewise,} \end{cases}$$
$$E_k = (c_1 + ak)2^k + (c_2 + c_3 ak)(2^k)^{\frac{1}{2}} + c_4$$
$$\lim_{k\to\infty} \frac{E_k}{k2^k} = a \Rightarrow E_k \sim aN_k \log N_k, \quad k \to \infty$$

for some scalars c_i independent on k.

- *after k succesive newest vertex bisection refinement steps around the corner* $(0,0)$ *as shown in fig. 2.9*

$$N_k = \begin{cases} \frac{3}{2}k & \text{if } k \text{ even} \\ \frac{3}{2}k - \frac{1}{2} & \text{elsewise,} \end{cases}$$
$$E_k = \tfrac{3}{4}ak^2 + \tfrac{1}{4}(12b - a)k$$
$$\lim_{k\to\infty} \frac{E_k}{k^2} = \tfrac{3}{4}a \Rightarrow E_k \sim \tfrac{3}{4}aN_k^2, \quad k \to \infty.$$

Proof. As there exist positive scalars a and b such that $|C(\varphi)| \leq al(\varphi) + b$ for every node $\varphi \in V$ the number E_k is bounded by

$$\begin{aligned} E_k &= \sum_{i=1}^{N_k} |C(\varphi_i)| \leq \sum_{i=1}^{N_k} al(\varphi) + b \\ &= \sum_{i=1}^{k} (N_i - N_{i-1})(ai + b) + bN_0 \leq \sum_{i=1}^{k} (N_i - N_{i-1})(ak + b) + bN_0 \\ &= (ak + b)N_k - akN_0 \end{aligned}$$

leading to $|J| = O(kN_k)$, $k \to \infty$. Now N_k and E_k are determined for the two cases above.

Note that the number of vertices after k steps of uniform newest vertex bisection of the unit-square is equal to the number of vertices after $k/2$ steps of uniform regular refinement for k even (compare fig. 3.1 and fig. 3.2), so for k even the formula for N_k is straightforward to derive. For k odd the formula is obtained by taking differences from adjacent even levels. Then E_k is calculated exactly by taking the sum over the couplings sets as in (4.1). Note that for k even $k = {}^2\log((\sqrt{N_k} - 1)^2)$ leading to an asymptotic behaviour of $E_k \sim aN_k \log N_k$.

The formula for N_k and E_k for the refinement around the origin $(0,0)$ can easily be derived with the use of fig. 2.9.

∙

This theorem suggests that the upperbound $O(kN)$ will be somewhere between $O(N \cdot {}^2\log N)$ and $O(N^2)$ if the number of refinement levels k is not restricted. If it is bounded then the hierarchical matrix will have $O(N)$ entries. For the newest vertex refinement method $|C(\varphi)| \leq \frac{3}{2}l(\varphi) + 4$ for all $\varphi \in V - v_0$ and $|C(\varphi)| \leq 4$ for all $\varphi \in v_0$ due to the regular numbering of the coarse mesh leading to $a = \frac{3}{2}$ and $b = 4$ in the theorem above.

As far as the computation of an entry of the hierarchical matrix is concerned note that the matrix can be assembled node-wise in parallel. However, this can lead to memory bank conflicts because different nodes may have common ancestors. An entry related to a coupling of two nodes of equal level can be computed with the use of an element matrix as usual in finite elements, the fast factorization method proposed in [5] may be used to this end. An entry related to a coupling between nodes of different levels is always a linear combination of the smallest triangles element matrix entries.

Matrix vector multiplications can be performed with the use of the parent function P without the use of the map M and do not involve checks on J. A sparse incomplete Choleski factorization like that proposed in [11] is easy to construct for the special storage scheme.

5 The C.-B.-S. scalar for the Laplacian equation

In section 3 it has been shown that frequently used mesh refinement techniques lead to a hierarchical matrix with a structured sparsity pattern. However, in order to exploit this sparsity patterns structure for the construction of efficient multilevel preconditioners, the mesh refinement techniques considered must satisfy additional conditions. For the regular refinement method these criteria, related to the angles in the mesh under consideration, have already been studied in the literature (see below) but for the newest vertex bisection case very little seems to known.

For the sake of complection therefore, this section will quote well-known estimates for the regular refinement method and will derive analogous results for the newest vertex bisection case. Considered are angle bounds and the computation of the mesh geometry dependent Cauchy-Buniakowskii-Schwarz scalar γ^2. This latter scalar determines the rate of convergence of algebraic multilevel preconditioners as considered in [6] and [7]. As is well-known, the angle bounds are of great importance for the determination of the discretization error. In order to obtain reasonable discretization error estimates the angles should be bounded below and away from π.

First consider the role of the scalar γ^2. Discretizing a linear partial differential equation using finite element basisfunctions defined on an underlying mesh geometry leads (see section 6) to a system of linear equations

$$H_k \underline{x}_k = \underline{b}_k \tag{5.1}$$

where the matrix and vector representation depend on the type of finite element basis considered. Suppose a two-level hierarchical base (see [6], [7], [9] and [17] for multilevel methods) is used whence, due to the two-level block stucture, the system can be written as

$$H_k = \begin{bmatrix} A_{k-1} & H_{12,k} \\ H_{21,k} & H_{22,k} \end{bmatrix} = I_k^t A_k I_k \quad \underline{x}_k := \begin{bmatrix} \underline{x}_{1,k} \\ \underline{x}_{2,k} \end{bmatrix} \quad \underline{b}_k := \begin{bmatrix} \underline{b}_{1,k} \\ \underline{b}_{2,k} \end{bmatrix} \tag{5.2}$$

where

- A_k is the matrix representing the Laplacian operator on the finite element basis defined standard nodally on $\underline{V}_k$.
- H_k is the matrix representing the Laplacian operator on the finite element basis defined standard nodally on $\underline{V}_{k-1}$ and hierarchically on $\underline{v}_k$.
- I_k is the transformation matrix representing the Identity operator from the hierarchical to the standard nodal basis as defined above.
- $\underline{x}_k$ and $\underline{b}_k$ are the solution and data vector.

The solution of the system $H_k \underline{x}_k = \underline{b}_k$ above is split into two parts (see also [1], [3], [8] and [19]) by the static condensation of the hierarchically defined nodes

$$\begin{cases} S_{k-1}\underline{x}_{1,k} = \underline{b}_{1,k} - H_{12,k} H_{22,k}^{-1} \underline{x}_{2,k} \\ \underline{x}_{2,k} = H_{22,k}^{-1}(\underline{b}_{2,k} - H_{21,k}\underline{x}_{1,k}) \end{cases} \tag{5.3}$$

where the *Schurcomplement* $S_{k-1} = (A_{k-1} - H_{12,k}H_{22,k}^{-1}H_{21,k})$ is a dense matrix in general. The rate of convergence of a PCG method for the solution of the first equation in (5.4) with A_{k-1} as a preconditioner depends on the C.-B.-S. scalar $0 < \gamma^2 < 1$ determined by

$$(1-\gamma^2)(A_{k-1}\underline{x},\underline{x}) \leq (S_{k-1}\underline{x},\underline{x}) \leq (A_{k-1}\underline{x},\underline{x}) \quad \forall_{\underline{x}\in\mathbb{R}^{N_k},\underline{x}\in Ker^{\perp}(A_{k-1})}, \tag{5.4}$$

since the latter inequality is trivially satisfied (see [6]). Due to

$$(1-\gamma^2)A_{k-1}\underline{x} = S_{k-1}\underline{x} \Leftrightarrow (H_{12,k}H_{22,k}^{-1}H_{21,k} - \gamma^2 A_{k-1})\underline{x} = 0$$

the scalar γ^2 can be obtained using the relation

$$\gamma^2 = \max\{\lambda \in \mathbb{R}: \{0\} \neq Ker(H_{12,k}H_{22,k}^{-1}H_{21,k} - \lambda A_{k-1}) \subset Ker^{\perp}(A_{k-1})\}\,. \tag{5.5}$$

An upperbound is given by $\gamma^2 \leq \max_{\Delta\in T}\{\gamma_\Delta^2\}$, where γ_Δ^2 for all triangles $\Delta \in T$ is computed with the use of (5.5) (see e.g. [6]). The scalar γ^2 is in fact a measure of the cosine of the angle between the span of the nodal and hierarchical basis (see [3]). As it depends on the finite element basisfunctions used it will be denoted by γ_l^2 and γ_q^2 in the linear resp. quadratic case from now on. First γ^2 is considered for the well-known case of regular refinement, thereafter it will be computed for the bisection case. In order to simplify future notations (ζ, α, β) will denote a triangle with angles ζ, α and β oriented counter-clockwise and the first angle ζ corresponding to the newest vertex.

Lemma 5.1 *Consider a triangle $\Delta \in T$ with angles α, β and ζ. For the Laplace equation the regular refinement of this triangle in the linear resp. quadratic case leads to*

$$\gamma_l^2 = \frac{3}{4}\gamma_q^2, \quad \gamma_q^2 = \tfrac{1}{2} + \frac{1}{3}\sqrt{d - \frac{3}{4}} \tag{5.6}$$

where $d = \cos^2\alpha + \cos^2\beta + \cos^2\zeta$.
Proof. See [19] for the linear and the quadratic case. Note that in the linear case γ^2 always is bounded away from one.

•

Theorem 5.1 *Consider a triangle $\Delta \in T$ with angles α, β and ζ the angle to be bisected. For the Laplace equation the newest vertex bisection of this triangle in the linear resp. quadratic case leads to*

$$\gamma_l^2 = \tfrac{1}{2}\frac{(\tilde\alpha + \tilde\beta)^2}{2 + \tilde\alpha^2 + \tilde\beta^2}, \quad \gamma_q^2 = \max\{\tfrac{1}{2}, \gamma_l^2, \gamma_c^2\} \tag{5.7}$$

where

$$\gamma_c^2 = \tfrac{1}{2}\frac{(\tilde\alpha+\tilde\beta)^4}{(\tilde\alpha^2+\tilde\beta^2+2)^2 + 4(\tilde\alpha^2+1)(\tilde\beta^2+1)} \cdot \frac{\tilde\alpha^2+\tilde\alpha\tilde\beta+\tilde\beta^2+3}{\tilde\alpha^2+\tilde\alpha\tilde\beta+\tilde\beta^2+1}$$

and $\tilde\alpha$, $\tilde\beta$ and $\tilde\zeta$ denote the cotangents of the corresponding angles.
Proof. Suppose that a triangle $\Delta \in T$ with nodes $\{\varphi_r\}_{r=1}^{p}$ is bisected such that q additional hierarchically defined nodes are created and define the corresponding hierarchical Laplacian element matrix H by

$$H = \begin{bmatrix} A & H_{12} \\ H_{21} & H_{22} \end{bmatrix} \tag{5.8}$$

where $H_{rs} := \int_\Delta \underline{\nabla}\varphi_s^t \underline{\nabla}\varphi_r$ for all $r, s = 1, \ldots, p+q$. Note that for a node defined linearly resp. quadratically on Δ its restriction to a child will also be linear resp. quadratic and hence is a linear combination of the childs nodes. Therefore all entries of H can be computed with the use of linear combinations of the standard nodally defined block A.

Further, suppose that the angle ζ is bisected into angles δ_1 and δ_2 such that children $(\delta_1, \alpha, \epsilon_1)$ and $(\delta_2, \varepsilon_2, \beta)$ are created (see fig. 1.4). Let the superscript '$\tilde{\ }$' denote the cotangent corresponding to each angle and note that

$$\tilde{\alpha} + \tilde{\beta} > 0 \quad \forall_{0<\alpha+\beta<\pi} \tag{5.9}$$

because the cotangent is a strictly decreasing function on $(0, \pi)$ with properties

$$\cot(\pi - x) = -\cot(x), \quad \cot(x+y) = \frac{\cot(x)\cot(y) - 1}{\cot(x) + \cot(y)} \quad \forall_{x,y \in (0,\pi)} . \tag{5.10}$$

From now on the linear and quadratic case will be considered seperately.

If linear basisfunctions are used the newest vertex bisection of ζ will create one new node, leading to the 4 by 4 matrix

$$H = \frac{1}{2} \begin{bmatrix} (\tilde{\alpha} + \tilde{\beta}) & -\tilde{\beta} & -\tilde{\alpha} & -(\tilde{\alpha} + \tilde{\beta}) \\ -\tilde{\beta} & (\tilde{\beta} + \tilde{\zeta}) & -\tilde{\zeta} & \tilde{\beta} \\ -\tilde{\alpha} & -\tilde{\zeta} & (\tilde{\zeta} + \tilde{\alpha}) & \tilde{\alpha} \\ -(\tilde{\alpha} + \tilde{\beta}) & \tilde{\beta} & \tilde{\alpha} & (\tilde{\alpha} + \tilde{\beta} + \tilde{\delta}_1 + \tilde{\delta}_2) \end{bmatrix} .$$

Here the rank of H_{22} equals one, which implies the existence of exactly one generalized eigenvalue λ satifying the condition above. Because matrix A has exactly one eigenvector $[1,1,1]^t$ corresponding to the eigenvalue 0, the condition'$\subset Ker^{\perp}(A)$' requires the transformation of $H_{12}H_{22}^{-1}H_{21} - \lambda A$ onto a basis which enables the elimination of this eigenvector. Then $T^{-1}(H_{12}H_{22}^{-1}H_{21} - \lambda A)T$, the transformed problem, is solved, where in this case the orthonormal matrix

$$T = \begin{bmatrix} \frac{1}{\sqrt{3}} & \frac{1}{\sqrt{2}} & \frac{1}{\sqrt{6}} \\ \frac{1}{\sqrt{3}} & 0 & -\frac{2}{\sqrt{6}} \\ \frac{1}{\sqrt{3}} & -\frac{1}{\sqrt{2}} & \frac{1}{\sqrt{6}} \end{bmatrix}$$

transforms the standard basis to the basis $((1,1,1),(0,1,0),(0,0,1))$. Note that, by construction of this transformation, the transformed matrix has first row and column equal to zero, whence a 3×3 minor system is left to be solved. This system obviously must have two eigenvalues equal to zero, because the rank of H_{22} is equal to one, and it appears to have a third generalized eigenvalue

$$\gamma_l^2 = \frac{\tilde{\alpha} + \tilde{\beta}}{\tilde{\alpha} + \tilde{\beta} + \tilde{\delta}_1 + \tilde{\delta}_2} .$$

In order to check whether $0 < \gamma_l^2 < 1$, the relationships (see also Mitchell [21])

$$\cot(\varepsilon_1) = \frac{\sin(\alpha - \beta)}{2\sin(\alpha)\sin(\beta)}, \quad \varepsilon_2 = \pi - \varepsilon_1$$

lead by (5.10) to

$$\tilde{\varepsilon}_1 + \tilde{\varepsilon}_2 = 0$$

$$\tilde{\varepsilon}_1 = \tfrac{1}{2}(\tilde{\beta} - \tilde{\alpha}) \wedge \tilde{\varepsilon}_2 = \tfrac{1}{2}(\tilde{\alpha} - \tilde{\beta})$$

$$\tilde{\delta}_1 = \cot(\delta_1) = -\cot(\alpha + \varepsilon_1) = -\frac{\tilde{\alpha}\tilde{\varepsilon}_1 - 1}{\tilde{\alpha} + \tilde{\varepsilon}_1} = \frac{2 - \tilde{\alpha}(\tilde{\beta} - \tilde{\alpha})}{\tilde{\alpha} + \tilde{\beta}} \wedge \tilde{\delta}_2 = \frac{2 - \tilde{\beta}(\tilde{\alpha} - \tilde{\beta})}{\tilde{\alpha} + \tilde{\beta}}$$

$$\tilde{\delta}_1 + \tilde{\delta}_2 = \frac{4 + (\tilde{\alpha} - \tilde{\beta})^2}{\tilde{\alpha} + \tilde{\beta}}, \quad \tilde{\delta}_1 - \tilde{\delta}_2 = \tilde{\alpha} - \tilde{\beta} .$$

The substitution of these latter relations into the previously obtained formula for γ_l^2 will yield the formula as given in the theorem.

In the case of quadratic basisfunctions defined on the three vertices and the midpoints of the three edges of the parent, the newest vertex bisection leads to the creation of three new nodes (see fig. 1.1). Here H is the 9×9 symmetric matrix with blocks A (see [2]), H_{22} given by

$$\frac{1}{3}\begin{bmatrix} 3(\tilde\alpha+\tilde\beta) & \tilde\beta & \tilde\alpha & 0 & -4\tilde\alpha & -4\tilde\beta \\ \tilde\beta & 3(\tilde\beta+\tilde\zeta) & \tilde\zeta & -4\tilde\zeta & 0 & -4\tilde\beta \\ \tilde\alpha & \tilde\zeta & 3(\tilde\zeta+\tilde\alpha) & -4\tilde\zeta & -4\tilde\alpha & 0 \\ 0 & -4\tilde\zeta & -4\tilde\zeta & 8d_0 & -8\tilde\beta & -8\tilde\alpha \\ -4\tilde\alpha & 0 & -4\tilde\alpha & -8\tilde\beta & 8d_0 & -8\tilde\zeta \\ -4\tilde\beta & -4\tilde\beta & 0 & -8\tilde\alpha & -8\tilde\zeta & 8d_0 \end{bmatrix}$$

$$\frac{8}{3}\begin{bmatrix} d_1+d_2 & -\tilde\varepsilon_1 & -\tilde\varepsilon_2 \\ -\tilde\varepsilon_1 & d_1 & 0 \\ -\tilde\varepsilon_2 & 0 & d_2 \end{bmatrix}$$

and $H_{12} = H_{21}^t$ defined by

$$\frac{1}{3}\begin{bmatrix} -3(\tilde\alpha+\tilde\beta) & 0 & 0 \\ -(\tilde\alpha+\tilde\beta+\tilde\delta_1+\tilde\delta_2+4\tilde\varepsilon_1) & 3\tilde\alpha-\tilde\delta_1+4\tilde\varepsilon_1 & -(\tilde\beta+\tilde\delta_2) \\ -(\tilde\alpha+\tilde\beta+\tilde\delta_1+\tilde\delta_2+4\tilde\varepsilon_2) & -(\tilde\alpha+\tilde\delta_1) & 3\tilde\beta-\tilde\delta_2+4\tilde\varepsilon_2 \\ -2(\tilde\alpha+\tilde\beta-\tilde\delta_1-\tilde\delta_2+2\tilde\varepsilon_1+2\tilde\varepsilon_2) & 6\tilde\alpha+3\tilde\delta_1+6\tilde\varepsilon_1 & 6\tilde\beta+3\tilde\delta_2+6\tilde\varepsilon_2 \\ 4(\tilde\alpha+\tilde\beta+\tilde\delta_1-\tilde\delta_2+\tilde\varepsilon_1+\tilde\varepsilon_2) & -4\tilde\varepsilon_1 & -8\tilde\beta+4\tilde\varepsilon_2 \\ 4(\tilde\alpha+\tilde\beta-\tilde\delta_1+\tilde\delta_2+\tilde\varepsilon_1+\tilde\varepsilon_2) & -8\tilde\alpha+4\tilde\varepsilon_1 & -4\tilde\varepsilon_2 \end{bmatrix} \tag{5.11}$$

where $d_0 = \tilde\alpha+\tilde\beta+\tilde\zeta$, $d_1 = \tilde\alpha+\tilde\delta_1+\tilde\varepsilon_1$ and $d_2 = \tilde\beta+\tilde\delta_2+\tilde\varepsilon_2$. The substitution of the expressions derived for $\tilde\varepsilon_1, \tilde\varepsilon_2, \tilde\delta_1, \tilde\delta_2$ into the elements of the 3 by 3 block H_{22} leads to

$$Det(H_{22}) = (\frac{8}{3})^3(\tilde\alpha+\tilde\beta)^{-3}(\tilde\alpha^4+\tilde\beta^4+6\tilde\alpha^2\tilde\beta^2+8(1+\tilde\alpha^2+\tilde\beta^2))(2+\tilde\alpha^2+\tilde\beta^2)$$

showing that this block is invertible and of rank 3 for all angles $0 < \alpha+\beta < \pi$, whence there may be at most three generalized eigenvalues λ which satisfy $H_{12}H_{22}^{-1}H_{21} - \lambda A = 0$. Matrix A has an eigenvector $[1,1,1,1,1,1]'$ corresponding to eigenvalue 0, so in this case T will transform the standard basis into the basis $((1,1,1,1,1,1), (0,1,0,0,0,0), \ldots, (0,0,0,0,0,1))$. Then $T^{-1}(H_{12}H_{22}^{-1}H_{21} - \lambda A)T$ is a 6×6 matrix with first row and column equal to zero, so λ can be determined by computing the determinant of the remaining 5×5 minor, factorizing it, setting it equal to 0 and solving the equation thus obtained for λ. This yields two eigenvalues 0, as to be expected, and exactly the three positive eigenvalues mentioned in (5.7).

•

In contrast to the regular refinement, where the scalar γ^2 is a function of the sum of the squares of the cosines of the three angles of the triangle, in the newest vertex bisection case this is not the case, as is shown by

Theorem 5.2 *For the newest vertex bisection refinement of a triangle with an angle* $\zeta \in (0,\pi)$ *to be bisected*

(i) γ_l^2 *and* γ_c^2 *do not depend on the parameter* $d = \cos^2\alpha + \cos^2\beta + \cos^2\zeta$ *as is the case for the regular refinement.*

(ii) γ_l^2 *is a function of* ζ *satifying*

$$0 < \min\{\tfrac{1}{2}, \sin^2(\tfrac{1}{2}\zeta)\} \le \gamma_l^2(\zeta) \le \max\{\tfrac{1}{2}, \sin^2(\tfrac{1}{2}\zeta)\} < 1 \quad \forall_{\zeta\in(0,\pi)}$$

and hence is bounded away from 0 *and* 1 *independent on the other angles.*

(iii) γ_c^2 *is bounded away from* 1 *by*

$$0 < \gamma_c^2 < \frac{3c^2}{3c^2+4c+1} < 1 \quad \forall_{0<\alpha+\beta<\pi}$$

where $c = \frac{1}{4}(\cot(\alpha)+\cot(\beta))^2$.

Proof. First consider (i). Note that for a triangle $(\frac{1}{6}\pi, \frac{1}{3}\pi, \frac{1}{2}\pi)$ $d = 1$ and $\gamma_l^2 = \frac{1}{14}$ but that for $(\frac{1}{3}\pi, \frac{1}{2}\pi, \frac{1}{6}\pi)$ $d = 1$ and $\gamma_l^2 = \frac{3}{10}$. In the quadratic case for the triangle $(\frac{1}{2}\pi, \frac{1}{4}\pi, \frac{1}{4}\pi)$ $d = 1$ and $\gamma_c^2 = \frac{3}{8}$ but for $(\frac{1}{2}\pi, \frac{1}{3}\pi, \frac{1}{6}\pi)$ $d = 1$ and $\gamma_c^2 = \frac{11}{28}$.

Now consider (ii). For each triangle (ζ, α, β) with $\zeta \in (0, \pi)$ and $\alpha \leq \beta$ there exists a positive angle $\xi \in [0, \frac{1}{2}(\pi - \zeta))$ such that

$$\alpha = \frac{1}{2}(\pi - \zeta) - \xi \text{ and } \beta = \frac{1}{2}(\pi - \zeta) + \xi .$$

Here ξ is a measure for the deformation of the triangle, which will be isosceles for $\xi = 0$ and degenerated to a line when ξ reaches $\frac{1}{2}(\pi - \zeta)$. For $\alpha > \beta$ reverse the roles of α and β.

The substitution of the relations $\tilde{\alpha} = \cot(\frac{1}{2}(\pi - \zeta) - \xi)$ and $\tilde{\beta} = \cot(\frac{1}{2}(\pi - \zeta) + \xi)$ into (5.7) and setting

$$\begin{cases} \zeta \in (0, \pi) & x = \cos^2(\frac{1}{2}(\pi - \zeta)) \quad \Rightarrow x \in (0, 1) \\ \xi \in [0, \frac{1}{2}(\pi - \zeta)) & y = \cos^2(\xi) \quad \Rightarrow y \in (x, 1) \end{cases}$$

yields the formula

$$\gamma_l^2 = \frac{x(1-x)}{x + y(1 - 2x)} \quad \forall_{x \in (0,1), y \in (x,1)} .$$

An investigation of this latter formula in order to determine its minimum and maximum for a given x shows that for all $x \in (0, 1)$ the denominator will be positive and a linear function of y. For $x \in (0, \frac{1}{2})$, γ_l^2 will be a decreasing function of y, whereas for $x \in (\frac{1}{2}, 1)$ γ_l^2 will be an increasing function with respect to this argument. For $x = \frac{1}{2}$ independently of y, $\gamma_l^2 = \frac{1}{2}$. Therefore, in both former cases the bounds are provided by the extreme values $y = x$ and $y = 1$ corresponding to a triangle degerated to a line and an isosceles triangle, leading to

$$\min\{\tfrac{1}{2}, x\} \leq \gamma_l^2 \leq \max\{\tfrac{1}{2}, x\} .$$

Finally the substitution of $x = \cos^2(\frac{1}{2}(\pi - \zeta)) = \sin^2(\frac{1}{2}\zeta)$ yields the desired result. In the special case where $\zeta = \frac{1}{2}\pi$

$$\gamma_l^2(\tfrac{1}{2}\pi) = \tfrac{1}{2} \quad \forall_{0 < \alpha + \beta < \pi}$$

for all other angles α and β.

For a proof of (iii) note that the formulas in (5.7) are symmetrically in $\tilde{\alpha}$ and $\tilde{\beta}$. In order to eliminate the cross terms $\tilde{\alpha}\tilde{\beta}$ in these formulas substitute (a transformation to the axis of symmetry $\tilde{\alpha} - \tilde{\beta} = 0$)

$$\begin{bmatrix} \tilde{\alpha} \\ \tilde{\beta} \end{bmatrix} = \sqrt{c}\begin{bmatrix} 1 \\ 1 \end{bmatrix} + \sqrt{x}\begin{bmatrix} 1 \\ -1 \end{bmatrix} \text{ with } \begin{bmatrix} c \\ x \end{bmatrix} = \frac{1}{4}\begin{bmatrix} (\tilde{\alpha} + \tilde{\beta})^2 \\ (\tilde{\alpha} - \tilde{\beta})^2 \end{bmatrix}$$

yielding for all $x, c \in [0, \infty)$

$$0 < \gamma_l^2(\tilde{\alpha}, \tilde{\beta}) = \frac{c}{1 + x + c} \leq \frac{c}{1 + c} < 1$$

$$0 < \gamma_c^2(\tilde{\alpha}, \tilde{\beta}) = \frac{(3 + x + 3c)c^2}{(1 + x + 3c)(1 + 2x + x^2 + 2c + c^2)} \leq \frac{3c^2}{3c^2 + 4c + 1} < 1$$

because both γ_l^2 and γ_c^2 are decreasing functions of x on $[0, \infty)$ for fixed c, which attain their maximum at $x = 0$.

•

The scalars γ_Δ^2 now can be computed for arbitrary $\Delta \in T$ leading to a bound for the C.-B.-S. scalar γ^2 as below (5.5). For the newest vertex bisection refinement method the maximum over all triangles can be replaced by the maximum over $4T_0$ number of triangles due to the fact that each coarse mesh triangle has exactly four similarity classes of children, as is shown in fig. 1.3 (fig. 1.2 shows the congruent children for the first red regular refinement). For an arbitrary $\Delta \in T_0$ given by (ζ, α, β) these four classes of children defined by eight different angles are (see fig. 1.4)

$$(\zeta, \alpha, \beta),\ (\varepsilon_1, \delta_1, \alpha),\ (\varepsilon_2, \beta, \delta_2),\ (\pi - \zeta, \delta_2, \delta_1)\,.$$

The C.-B.-S. scalar upperbound therefore can be computed for an arbitrary refinement by considering only the coarse mesh triangles, see section 6 for an example. Due to the similarity class property also the angle bounds only depend on the coarse mesh geometry whence a properly chosen coarse mesh automatically avoids bad angles.

6 Numerical examples

In order to show that the sparsity pattern is structured in the case of local mesh refinement, two examples are studied for the unit-square $\Omega = (0,1)^2$. Thereafter the C.-B.-S. scalar is computed for some example triangles.

First consider the computation of the solution of the Laplacian problem: find $\hat{u} \in H_g^1(\Omega)$ such that

$$\begin{aligned} -\Delta \hat{u} &= f \text{ in } \Omega \\ \hat{u} &= g \text{ on } \Gamma \end{aligned} \tag{6.1}$$

where by definition
- $H^1(\Omega)$ is the Sobolev one Space of functions defined on Ω.
- $H_g^1(\Omega) = \{v \in H^1(\Omega) : v = g \text{ on } \Gamma\}$ for a given trace g on Γ.
- f is a square integrable source function in Ω.

The equation is discretized by choosing an initial coarse mesh, denoted by Ω_0, finite element basisfunctions thereon, and a mesh refinement method. Intermediate levels of refinement correspond to unique meshes Ω_k defined by the sets T_k as defined in section 2. If a mesh Ω_k has been refined, a function in W_k, the nodal span of V_k, is interpolated to the finer mesh Ω_{k+1} as follows. For a function $u \in H_g^1(\Omega_k) := H_g^1(\Omega) \cap W_k$ its interpolant $G(u) \in W_{k+1}$ is uniquely determined by its values at the vertices of $\underline{V}_{k+1}$ given by

$$\begin{cases} (G(u))(\hat{\varphi}) = & u(\hat{\varphi}) \quad \text{if } \hat{\varphi} \in \underline{V}_{k+1} \cap \underline{V}_k \\ & u(\hat{\varphi}) \quad \text{if } \hat{\varphi} \in \underline{V}_{k+1} - \underline{V}_k \wedge \hat{\varphi} \in \Omega \\ & g(\hat{\varphi}) \quad \text{if } \hat{\varphi} \in \underline{V}_{k+1} - \underline{V}_k \wedge \hat{\varphi} \in \Gamma \end{cases}$$

independent on the finite element basis used.

Provided an initial solution $\hat{u}_0 \in H_g^1(\Omega_0)$ on the coarse initial mesh, the discrete solution on the finest mesh will be computed with the use of approximations of the discrete solutions $\hat{u}_i \in H_g^1(\Omega_i)$, $0 < i < k$ on coarser meshes by

$$\begin{aligned} u_k &= G(\hat{u}_{k-1}) \\ H_k \underline{x}_k &= \underline{b}_k \\ \hat{u}_k &= u_k + x_k \end{aligned} \tag{6.2}$$

where H_k is the matrix representing the Laplacian operator on the unit-square, $\underline{b}_k = -(H_k\underline{u}_k - \underline{f}_k)$ and $\underline{f}_k$ the source function data determined by f. Here the vector representation $\underline{u}$ of a function u and the matrix H_k are basis dependend. The identity transformation I_k as defined in section 1 determines the relation between the hierarchical resp. standard nodal basis vector representation. Note that $\underline{x}_k \in \mathrm{H}^1_{0,h}(\Omega)$ by definition.

Now consider two example differential equations to be considered in more detail. The first one is given by

$$\begin{aligned} -\Delta u &= 0 \text{ in } \Omega \\ u &= g \text{ on } \Gamma \end{aligned} \tag{6.3}$$

for smooth Dirichlet boundary data g. Here the unit-square is refined using uniform newest vertex mesh refinement, i.e., Ω_{k+1} is constructed from Ω_k by refining all triangles in $\tau_k \in T_k$. The inital coarse mesh Ω_0 is as in the first picture of fig. 3.1. The second example is the convection-diffusion equation

$$\begin{cases} -\varepsilon u_{xx} + 2u_x + u_y = 0 & 0 < x < 1, 0 < y \leq 1 \\ u(0,y) = 1 & 0 \leq y \leq 1 \\ u(1,y) = 0 & 0 \leq y \leq 1 \\ u(x,0) = 0 & 0 < x < 1\,. \end{cases} \tag{6.4}$$

The solution of this equation has a shock along the line $y = \frac{1}{2}x$ whence Ω_{k+1} is constructed from Ω_k by refining those triangles intersecting this line of level smaller than $k+1$ without children. Like in the previous example newest vertex mesh refinement is used starting from the same initial coarse mesh. For both examples the sparsity patterns are computed for the hierarchical and for the standard nodal finite element basis.

k	$\lvert\tau_k\rvert$	$\lvert T_k\rvert$	n_k	N_k	E_k (H)	E_k (N)
0	2	2	4	4	14	14
1	4	6	1	5	23	21
2	8	14	4	9	59	41
3	16	30	4	13	115	69
4	32	62	12	25	279	137
5	64	126	16	41	575	249
6	128	254	40	81	1319	497
7	256	510	64	145	2775	945
8	512	1022	144	289	6151	1889
9	1024	2046	256	545	13031	3681

Table 6.1 The uniformly refined unit-square.

For the first example fig. 4.1 shows Ω_7 together with the numbering of the nodes. Then fig. 4.2 and fig. 4.4 present the sparsity pattern for a hierarchical resp. standard nodal basis on this mesh geometry and figs. 4.3, 4.5 show the patterns' subset corresponding to non-zero matrix entries. Black dots are plotted in these pictures for those $(i,j) \in J$, satisfying the conditions on h_{ij}, in such a way that $(1,1)$ is situated in the top-left corner and (N,N) in the bottem-right corner. The horizontal and vertical lines show the block structure of the sparsity pattern and hierarchical matrix H, only serving reference purposes in the standard nodal case.

In the hierarchical case many entries are zero as is shown in fig. 4.3 and proved in [20] and some offdiagonal entries are positive contrary to the standard nodal case. Fig. 4.1 shows that the couplings set of a standard nodally defined node will contain at most five elements which one can prove to be valid if every node of level zero satisfies this assumption. Hence the substitution of 4 by 5 and of -3 by -4 in the special row-wise ordered storage scheme presented in section 4 enables additionally the storage of the standard nodal sparsity pattern. In the standard nodal case all nodes have level 0 by definition but still correspond to vertices of different levels. Fig. 4.5 shows that the upper left minor of the matrix tends to become diagonal for $k \to \infty$, since due to the uniform refinement nodes of lower level get de-coupled.

Table 6.1 gives for Ω_0 to Ω_9 an overview of the elementary parameters, all defined in section 2 to section 4. The parameters depending on the type of finite element basis used are followed by '(H)' or '(N)' to distuingish between the hierarchical resp. standard nodal base.

k	$\|\tau_k\|$	$\|T_k\|$	n_k	$n_{k,18}$	N_k	E_k(H)	E_k(N)
0	2	2	4	4	4	14	14
1	4	6	1	1	5	23	21
2	6	12	3	4	8	50	36
3	12	26	3	4	12	100	62
4	14	40	5	9	17	169	91
6	30	110	9	18	37	521	221
8	54	260	15	36	77	1395	491
12	198	1232	51	114	325	8537	2207
16	774	5228	195	449	1329	–	9215
18	1542	10586	387	387	2671	–	18599

Table 6.2 The locally refined unit-square.

Now consider the second example. Meshes Ω_{12} and Ω_{18} are shown in figs. 5.1 and 5.4. A magnification of the latter can be found in fig. 5.5 demonstrating that the newest vertex bisection refinement technique is capable of capturing small details departing from a coarse initial mesh. The sparsity pattern for Ω_{12} is plotted for a hierarchical resp. standard nodal basis in figs. 5.2 and 5.3, for Ω_{18} it is depicted for the standard nodal basis in fig. 5.6. In this latter example clearly the sparsity pattern is locally very irregular but globally structured. The number of triangles and other mesh parameters for this example are listed in table 6.2.

Due to the recursiveness of the bisection algorithm presented in section 2 the refinement of a triangle of level k can create several nodes of level $0 \leq i \leq k$. During the uniform refinement of Ω_k in the first example this is not the case, only nodes of level $k+1$ are created which means that the sets of nodes v_i, $0 \leq i \leq k$ are not altered after the construction of Ω_k. This is different for the second example where the sets v_i grow due to the refinement of triangles of level $k > i$ as is demonstrated in table 6.2 in the fifth column. Here $n_{k,18}$ denotes the number of nodes for given level n_k after 18 levels of refinement.

$\Delta \backslash \gamma_l^2, \gamma_c^2$	1st *Class*	2nd *Class*	3rd *Class*	4th *Class*
$(\frac{1}{2}\pi, \frac{1}{4}\pi, \frac{1}{4}\pi)$	$\frac{1}{2}, \frac{3}{8}$	$\frac{1}{2}, \frac{3}{8}$	$\frac{1}{2}, \frac{3}{8}$	$\frac{1}{2}, \frac{3}{8}$
$(\frac{1}{3}\pi, \frac{1}{3}\pi, \frac{1}{3}\pi)$	$\frac{1}{4}, \frac{1}{8}$	$\frac{1}{2}, \frac{11}{28}$	$\frac{1}{2}, \frac{11}{28}$	$\frac{3}{4}, \frac{27}{40}$
$(\frac{1}{12}\pi, \frac{2}{3}\pi, \frac{1}{4}\pi)$	$0.268, 0.00157$	$\frac{2}{5}, 0.303$	$\frac{3}{5}, 0.668$	$0.973, 0.977$
$(\frac{2}{3}\pi, \frac{1}{12}\pi, \frac{1}{4}\pi)$	$0.661, 0.680$	$0.882, 0.873$	$0.118, 0.305$	$0.337, 0.206$
$(\frac{1}{4}\pi, \frac{2}{3}\pi, \frac{1}{12}\pi)$	$0.306, 0.166$	$0.0820, 0.0130$	$0.918, 0.938$	$0.649, 0.774$

Table 6.3 Scalars γ_l^2 and γ_c^2 for the newest vertex bisection.

In order to get an impression of the values γ_Δ^2 for some triangles note that in the case of linear basisfunctions the four similarity classes mentioned in section 5 correspond to the following C.-B.-S. scalars

$$\begin{aligned}\gamma_1^2, \gamma_2^2, \gamma_3^2, \gamma_4^2 =& \tfrac{1}{2}\frac{(\tilde{\alpha}+\tilde{\beta})^2}{\tilde{\alpha}^2+\tilde{\beta}^2+2}, \frac{\tilde{\alpha}^2+1}{\tilde{\alpha}^2+\tilde{\beta}^2+2}, \frac{\tilde{\beta}^2+1}{\tilde{\alpha}^2+\tilde{\beta}^2+2}, \tfrac{1}{2}\frac{(\tilde{\alpha}-\tilde{\beta})^2+4}{\tilde{\alpha}^2+\tilde{\beta}^2+2} \\ \equiv& \gamma_1^2, \gamma_2^2, 1-\gamma_2^2, 1-\gamma_1^2\end{aligned}$$

according to (5.7). Table 6.3 gives γ_l^2 and γ_c^2 for some example triangles (ζ, α, β) with ζ the angle to be bisected, computed with the use of (5.7). Note that γ_q^2 can be found with the use of the table and the relationship $\gamma_q^2 = \max\{\frac{1}{2}\gamma_l^2, \gamma_c^2\}$. According to table 6.3 in the quadratic case it is best to have a largest angle of approximately $\frac{1}{2}\pi$ bisected which was already shown by lemma 5.1 for the linear case. Clearly the newest vertex bisection and regular refinement method are well suited for multilevel preconditioning methods which shows that the advantages of a simple sparsity pattern and multilevel preconditioning may be combined for the construction of a preconditioner which is considered in more detail in [20].

7 Conclusions

- The use of the sparsity pattern reduces the number of indirect addressing which occurs when computations are performed elementwise.
- If the number of refinement levels k is not bounded then the total number of possibly non-zero entries of a hierarchical matrix is bounded above by $O(kN)$. In practice $\log N \leq k \leq N$. If the number of refinement levels is restricted then the number of possibly non-zero entries will be $O(N)$ as in the standard nodal finite element matrix case.
- The techniques provided in this report to determine upperbounds for the lengths of couplings sets can easily be extended to all mesh refinement techniques mentioned in the introduction, for higher order basisfunctions and in the case of more space dimensions.
- The simple recursive newest vertex bisection technique is well suited for multilevel preconditioning as the C.-B.-S. scalar γ^2 which determines the rate of convergence is well bounded below 1 and only depends on the initial coarse mesh.

8 References

[1] Axelsson O., *An algebraic framework for multilevel methods,* internal report 8820 (October 1988), Department of Mathematics, University of Nijmegen, The Netherlands

[2] Axelsson O. and Barker V.A., *Finite Element Solution of Boundary Value Problems,* Academic Press, Orlando, Fl., 1984

[3] Axelsson O., and Gustafsson I., *Preconditioning and two-level multigrid methods of arbitrary degree of approximation,* Mathematics of Computation 40(1983), 219-242

[4] Axelsson O. and Maubach J., *A time-space finite element discretization technique for the calculation of the electromagnetic field in ferromagnetic materials,* Journal for Numerical Methods in Engineering 29(1989), 2085-2111

[5] Axelsson O. and Maubach J., *On the updating and assembly of the hessian matrix in finite element methods,* Computer Methods in Applied Mechanics and Engineering 71(1988),41-67

[6] Axelsson O. and Vassilevski P.S., *Algebraic multilevel preconditioning methods I,* Numer. Math. 56(1989), 157-177

[7] Axelsson O. and Vassilevski P.S., *Algebraic multilevel preconditiong methods II,* Siam J. Numer. Anal. 1990, to appear

[8] Bank R.E, and Dupont T.F., *Analysis of a two-level scheme for solving finite element equations,* Report CNA-159, Center for Numerical Analysis, University of Texas, Austin, 1980

[9] Bank R.E., Dupont T.F. and Yserentant H., *The hierachical basis multigrid method,* Preprint SC 87-1 (1987), Konrad-Zuse-Zentrum für Informationstechnik, Berlin

[10] Bank R.E. and Sherman A.H., *The use of adaptive grid refinement for badly behaved elliptic partial differential equations,* in Advances in Computer Methods for Partial Differential Equations II (Vichnevetsky R. and Stepleman R.S. eds.), IMACS, 1979, 33-39

[11] Bank R.E. and Smith R.K., *General sparse elimination requires no permanent integer storage,* SIAM J. Sci. Stat. Comput. 8(1987), 574-584

[12] Bank R.E. and Weiser A., *Some a posteriori error estimators for elliptic partial differential equations,* Math. Comp. 44(1985), 283-301

[13] Becker E.B., *Finite Elements,* Prenctice Hall, Englewood Cliffs, New Jersey, 1981

[14] Carey G.F., Barragy R., Mclay R. and Sharma M., *Element by element vector and parallel computations,* Communications in Applied Numerical Methods 4(1988), 299-307

[15] Carey G.F., Sharma M. and Wang K.C., *A class of data structures for 2-D and 3-D adaptive mesh refinement, in press,* International Journal for Numerical Methods in Engineering

[16] Ciarlet P.G., *The Finite Element Method for Elliptic Problems,* North-Holland Publ., Amsterdam, 1978

[17] Deuflhard P., Leinen P. and Yserentant H., *Concepts of an adaptive hierarchical finite element code,* Preprint SC 88-5 (September 1988), Konrad-Zuse-Zentrum für Informationstechnik, Berlin

[18] Greenbaum A., Congming Li and Han Zheng Chao, *Comparison of linear system solvers applied to diffusion-type finite element equations,* internal report Ultracomputer Note #126, New York University, September 1987

[19] Maitre J.F. and Musy F., *The contraction number of a class of two-level methods; an exact evalutation for some finite element subspaces and model problems,* in Multigrid Methods (Hackbusch W. and Trottenberg U. eds.), Proceedings Köln-Porz 1981, LNM 960, Springer Verlag, 1982, 535-544

[20] Maubach J.M., *On the sparsity pattern, block decay rates and preconditioning of multilevel hierarchical finite element matrices,* in preparation, intern. report, University of Nijmegen 1989

[21] Mitchell W.F., *A comparison of adaptive refinement techniques for elliptic problems,* internal report no. UIUCDCS-R-1375, Department of computer science, University of Illinois at Urbana-Champaign, Urbana, Illinois(1987), to appear in ACM Trans. Math. Soft.

[22] Mitchell W.F., *Unified multilevel adaptive finite element methods for elliptic problems,* internal report no. UIUCDCS-R-88-1436, Department of computer science, University of Illinois at Urbana-Champaign, Urbana, Illinois(1988)

[23] Rivara M.C., *Algorithms for refining triangular grids suitable for adaptive and multigrid techniques,* International Journal for Numerical Methods in Engeneering 20(1984), 745-756

[24] Rivara M.C., *A grid generator based on 4-triangles conforming mesh refinement algorithms,* International Journal for Numerical Methods in Engeneering 24(1987), 1343-1354

[25] Sewell E.G., *Automatic generation of triangulations for piecewise polynomial approximation,* Ph.D. Thesis, Purdue University, West Lafayette, IN, 1972

[26] Sewell E.G., *A finite element program with automatic user-controlled mesh grading,* in Advances in Computer Methods for Partial Differential Equations III (Vichnevetsky R. and Stepleman R.S. eds.), IMACS, 1979, 8-10

[27] Strang G. and Fix G.S., *An analysis of the finite element method,* Prentice Hall, Englewood Heights, New Jersey, 1973

[28] Yserentant H., *On the multilevel splitting of finite element spaces,* Numerische Mathematik 49(1986), 379-412

[29] Zienkiewicz O., *The Finite Element Method in Engineering Science,* 3rd edition, Mc Graw-Hill, New York, 1977

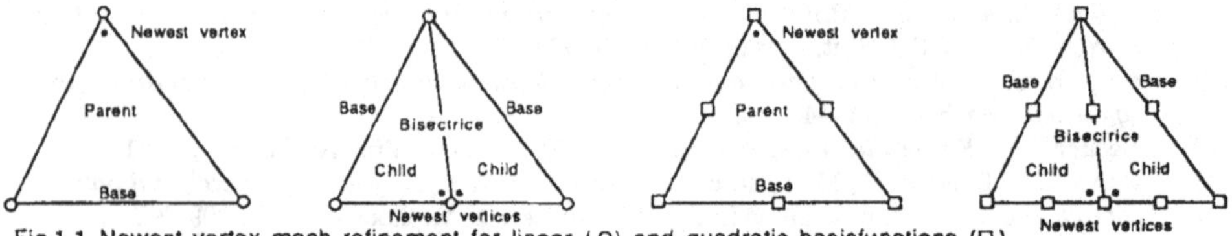

Fig.1.1 Newest vertex mesh refinement for linear (○) and quadratic basisfunctions (□).

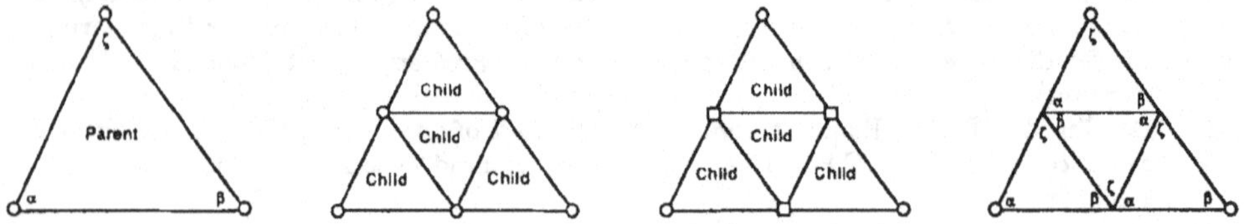

Fig.1.2 Regular red-1 mesh refinement for linear (○) and quadratic basisfunctions (□).

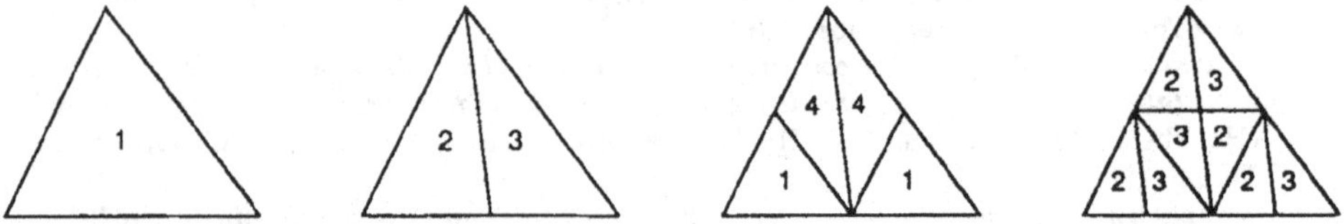

Fig.1.3 The four mesh refinement similarity classes.

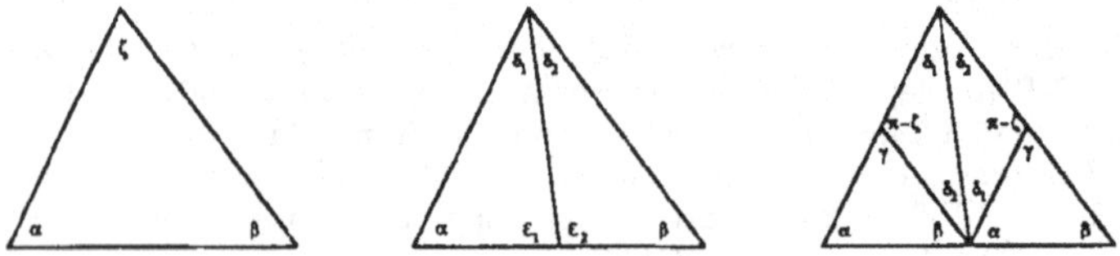

Fig.1.4 The eight angles involved with bisection.

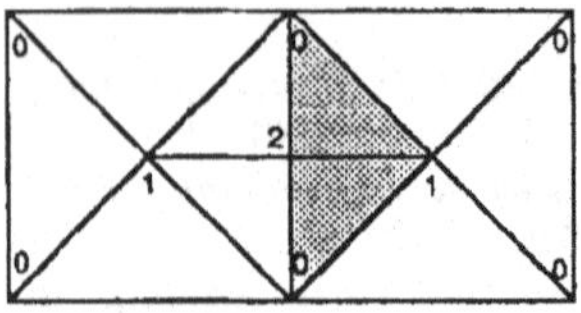

Fig.1.5 Level 2 newest vertex with a parent with two level 0 and one level 1 vertices.

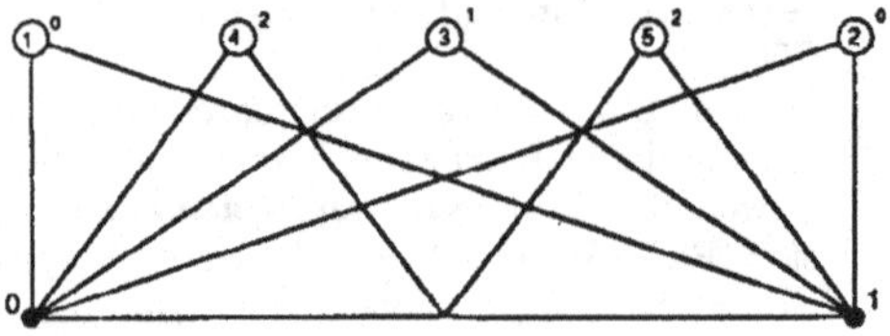

Fig.1.6 One dimensional hierarchical basisfunctions 1,2,3,4 and 5 with levels resp. 0,0,1,2 and 2.

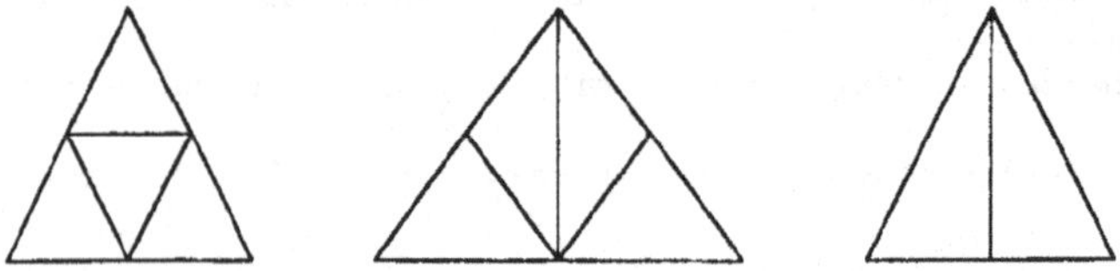

Fig.1.7 Regular red-1, red-2 and green type of refinement.

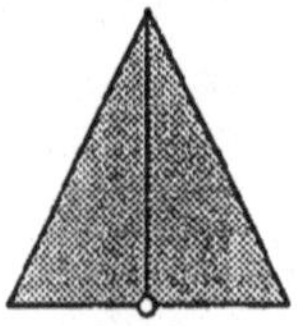

Fig.1.8 Domain of the node created by newest vertex bisection.

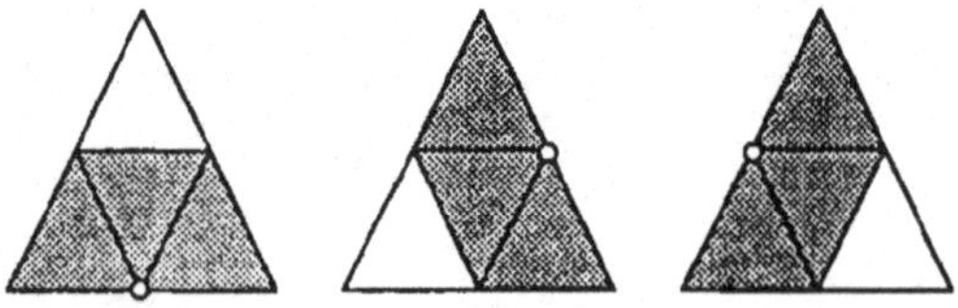

Fig.1.9 Domains of the nodes created by red-1 type regular refinement.

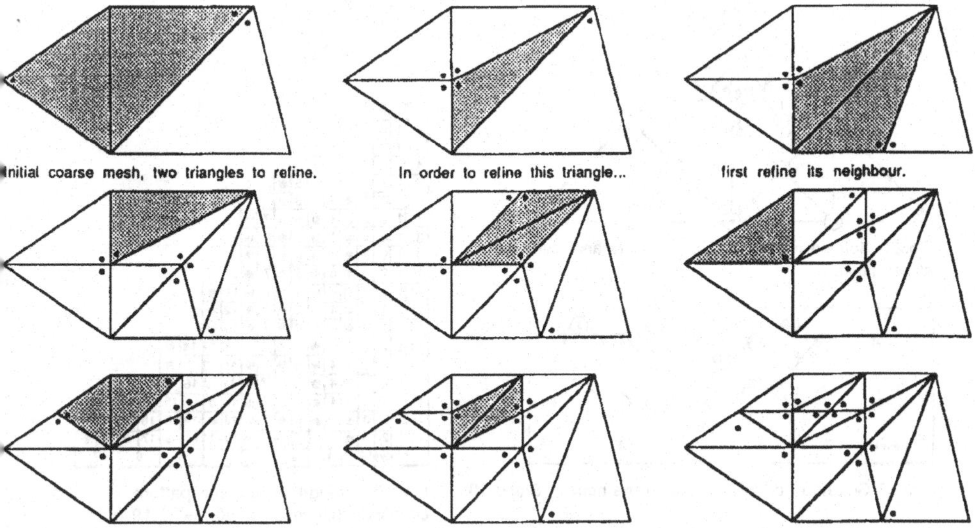

Fig.2.1 Step by step newest vertex mesh refinement.

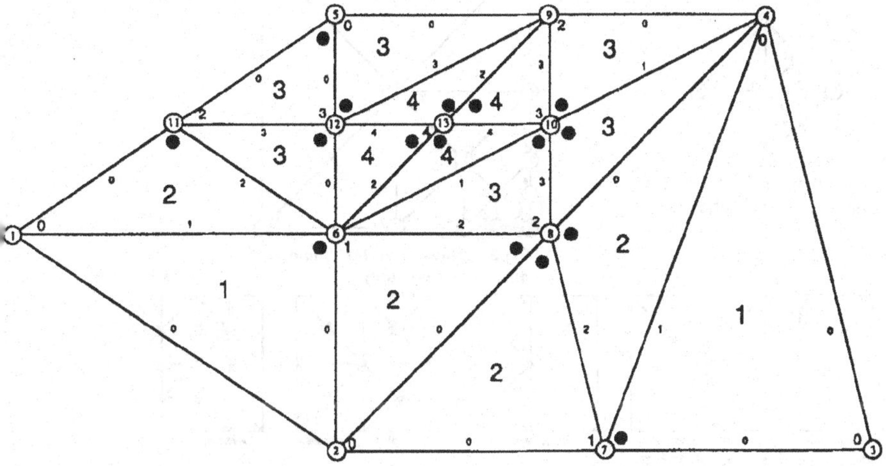

Fig.2.2 Levels of triangles, vertices, lines and numbers of vertices.

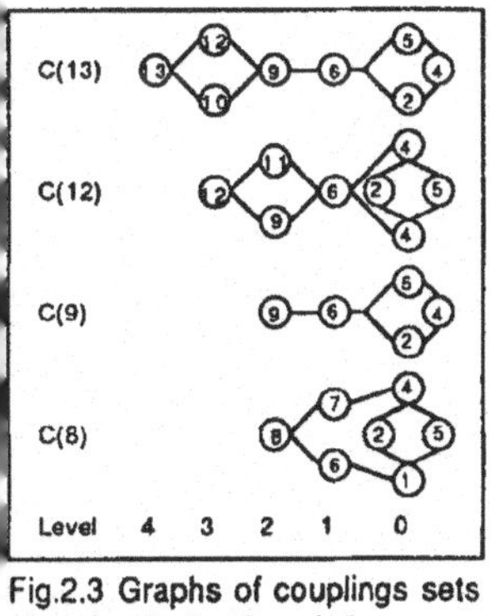

Fig.2.3 Graphs of couplings sets of nodes 8, 9, 12 and13.

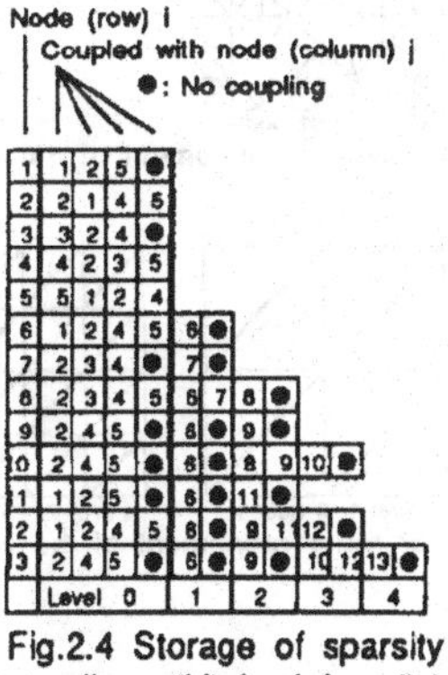

Fig.2.4 Storage of sparsity pattern couplings (i,j) in J for all i,j=1....13.

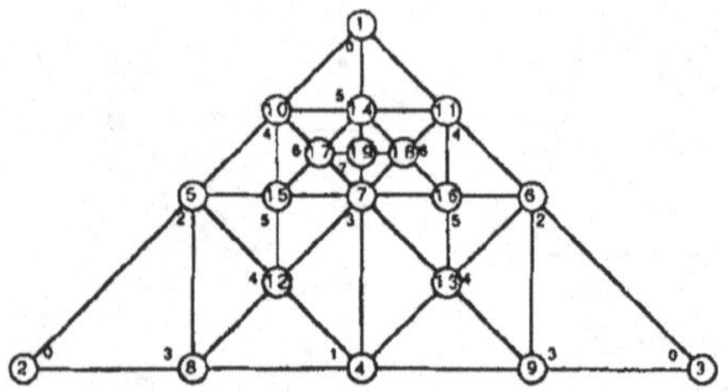

Fig.2.5 Refined mesh with the numbers and levels of all vertices.

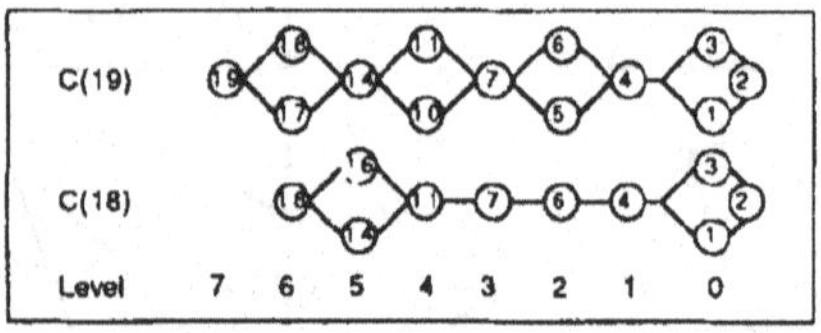

Fig.2.6 Graphs of couplings sets of the nodes 18 and 19.

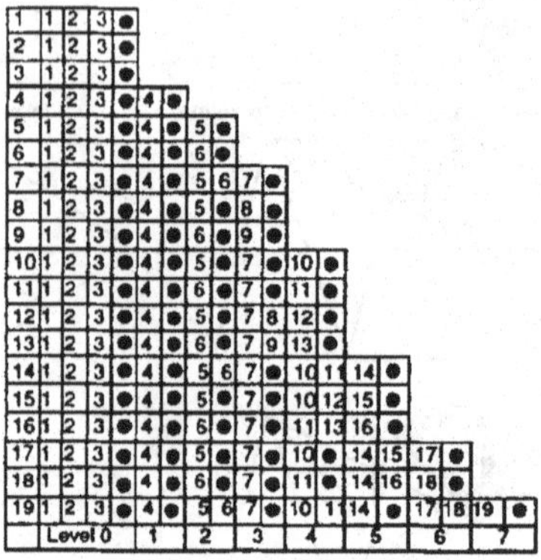

Fig.2.7 Storage of sparsity pattern couplings (i,j) in J for all i,j=1....19.

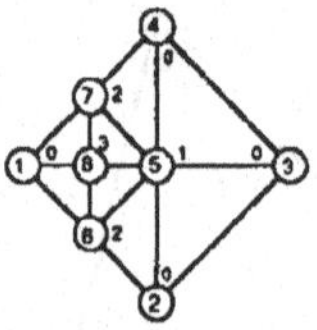

Fig.2.8 Nodes 7, 5 and 8 have different pathes.

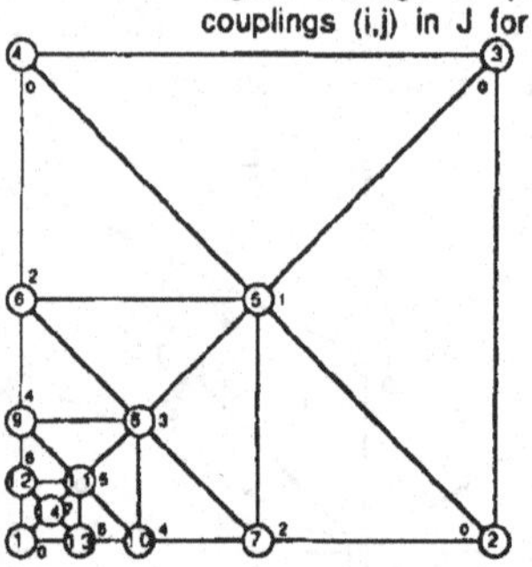

Fig.2.9 Newest vertex refinement around a singularity.

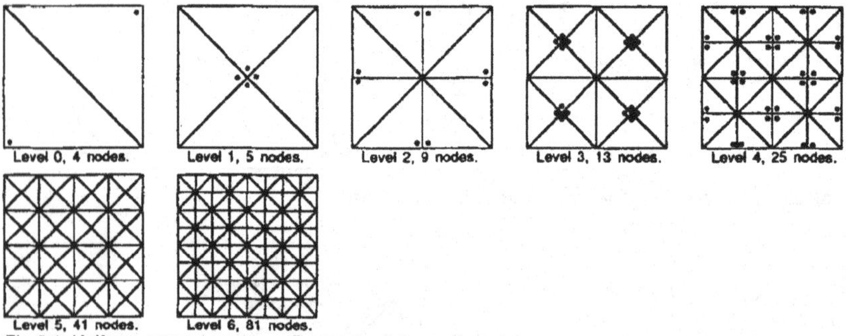

Level 0, 4 nodes. Level 1, 5 nodes. Level 2, 9 nodes. Level 3, 13 nodes. Level 4, 25 nodes.

Level 5, 41 nodes. Level 6, 81 nodes.

Fig.3.1 Uniform newest vertex refinement of the unit square.

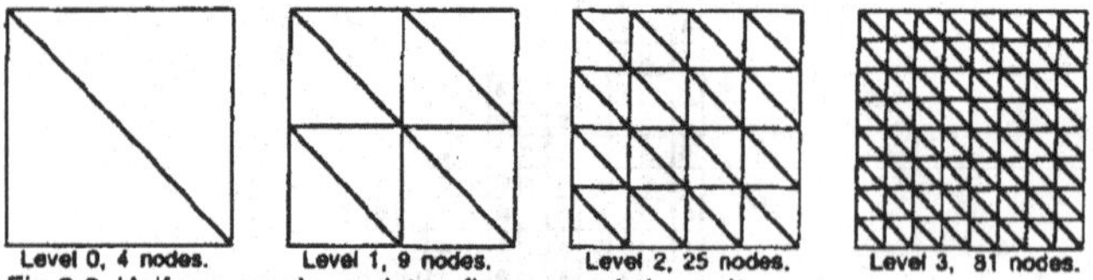

Level 0, 4 nodes. Level 1, 9 nodes. Level 2, 25 nodes. Level 3, 81 nodes.

Fig.3.2 Uniform regular red-1 refinement of the unit square.

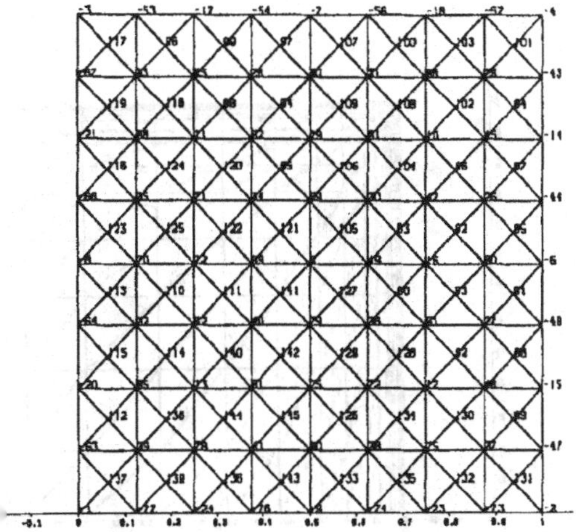

4.1 Uniform refinement, mesh Ω_7 with node numbers.

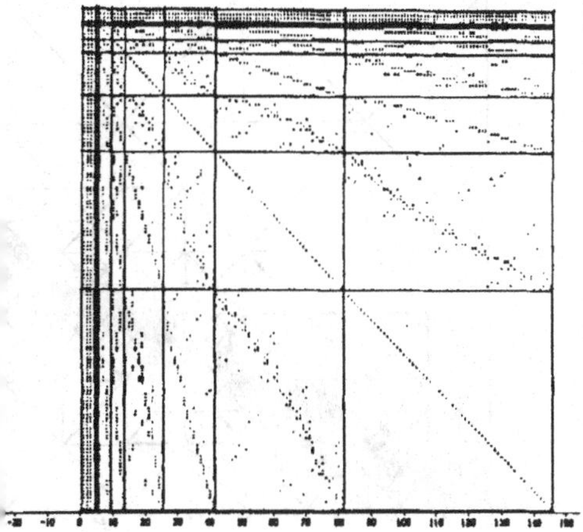

4.2 Uniform refinement, sparsity pattern (H).

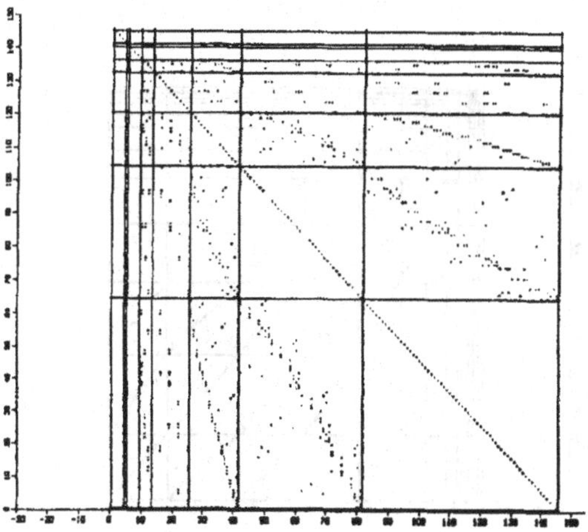

Fig. 4.3 Subset of couplings (i,j) with $h_{ij} \neq 0$ (H).

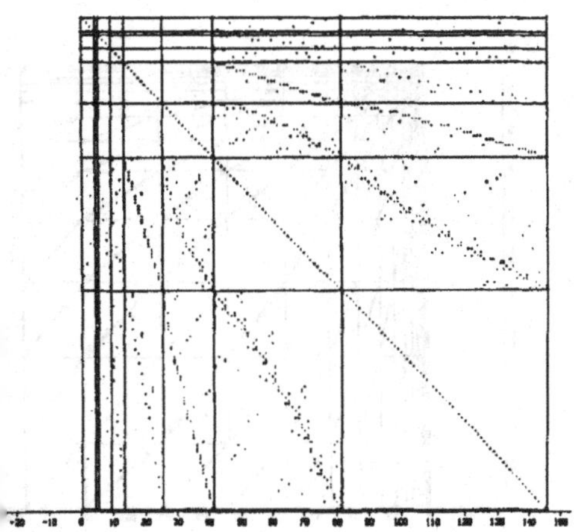

4.4 Uniform refinement, sparsity pattern (N).

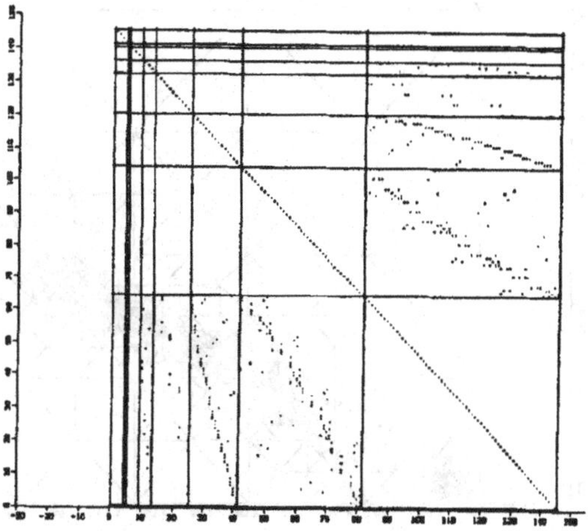

Fig. 4.5 Subset of couplings (i,j) with $h_{ij} \neq 0$ (N).

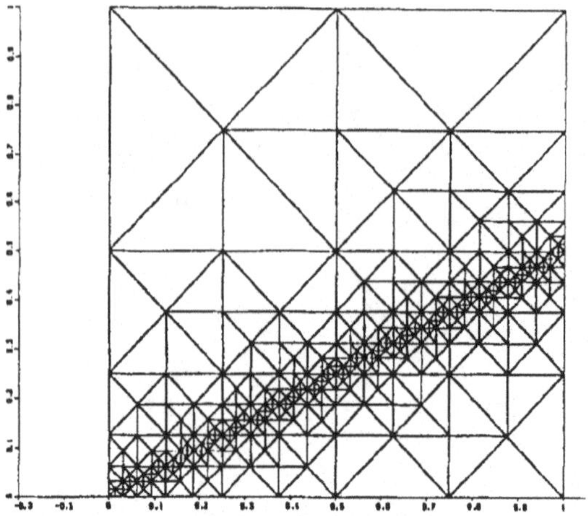

Fig. 5.1 Local refinement, mesh Ω_{12}.

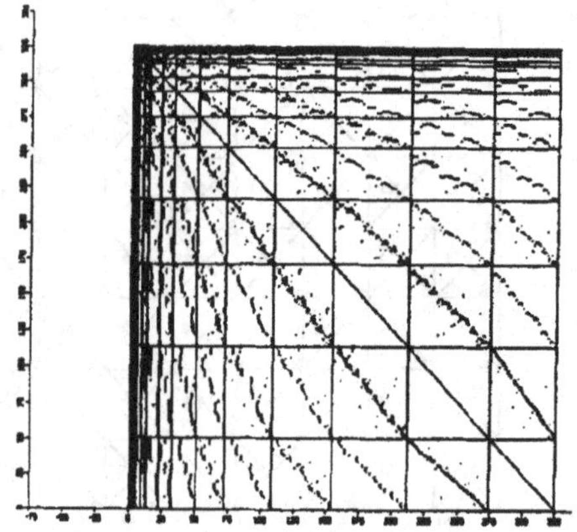

Fig. 5.2 Local refinement, sparsity pattern Ω_{12} (H)

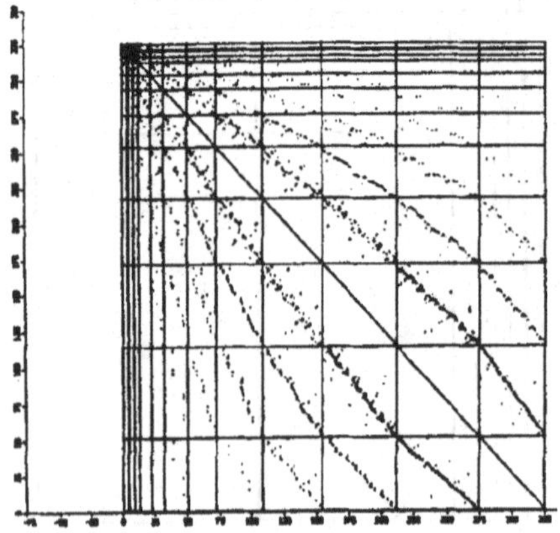

Fig. 5.3 Local refinement, sparsity pattern Ω_{12} (N).

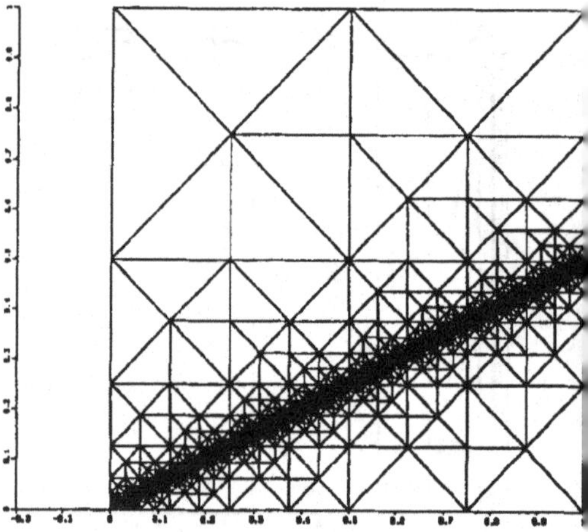

Fig. 5.4 Local refinement, mesh Ω_{18}.

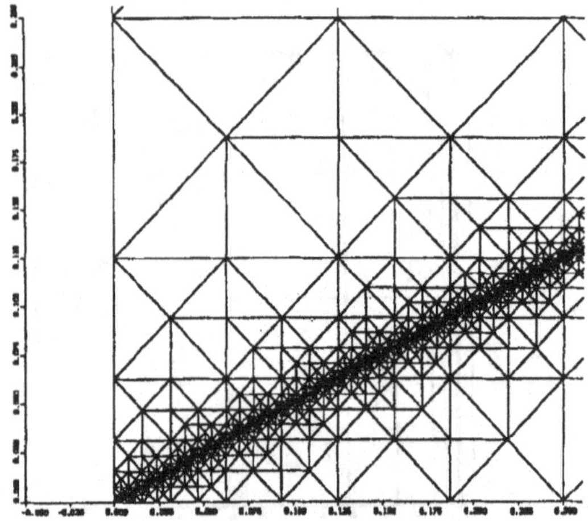

Fig. 5.5 Magnification of mesh Ω_{18}.

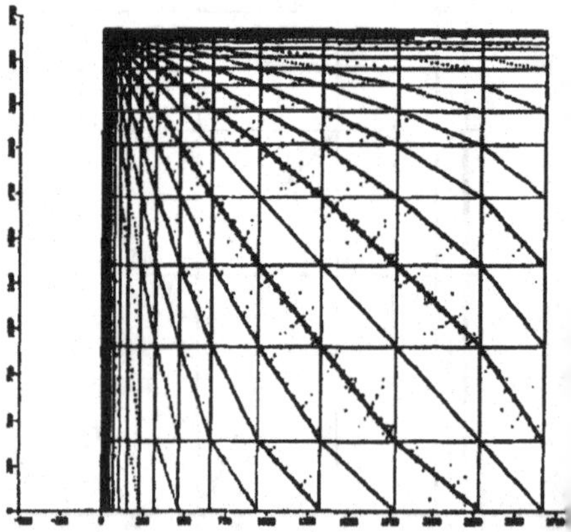

Fig. 5.6 Local refinement, sparsity pattern Ω_{18} (N

SOLVING POSITIVE (SEMI)DEFINITE LINEAR SYSTEMS BY PRECONDITIONED ITERATIVE METHODS

Y. NOTAY *

Abstract. The use of preconditionings obtained by so called modified incomplete factorizations has become quite popular for the PCG solution of regular systems arising from the discretization of elliptic PDE's. Our purpose here is to review their recent extension to the singular case. Because such conditionings may themselves be singular, we first review the extension of the general theory of polynomial acceleration to the case of singular preconditionings. We emphasize that all results can be formulated in such a way that they cover both the regular and singular cases. Examples of application are given, displaying the superiority of the recently developed factorization strategies.

Key Words. Iterative methods for linear systems, acceleration of convergence, preconditioning.

1. Introduction. Let A be a symmetric positive (semi)definite n x n matrix and consider the consistent system

$$Ax = b \tag{1.1}$$

with $b \in \mathcal{R}(A)$ (see below for notation). In the regular case, the PCG method with preconditioning arising from modified incomplete factorizations has become a popular technique and our purpose here is to review its recent extension to the singular case. Such an extension was not straightforward because the associated preconditioning matrix may itself be singular and no theory of singularly preconditioned polynomially accelerated iterative schemes had been developed. It has therefore been performed in two stages, both of which will be reviewed here :(1) the extension of the general theory of preconditioned polynomial iterative methods to the case of possibly singular preconditionings, cf. [23], and (2) the more specific theory of modified incomplete factorizations of possibly singular Stieltjes matrices, cf. [24]. We shall additionally stress that all results can be formulated in such a way that they cover both the regular and singular cases with, in the latter event, both regular and singular preconditioners.

We are therefore first interested here in preconditioned (polynomial) iterative methods for solving (1.1). That is we first make some splitting of the system matrix

$$A = B - C \tag{1.2}$$

where the "preconditioner" or "preconditioning matrix" B is symmetric nonnegative definite with

$$\mathcal{N}(B) \subset \mathcal{N}(A) \tag{1.3}$$

Then, (1.1) is solved by an iterative scheme of the following form, starting with an arbitrary initial approximation x_o :

* Université Libre de Bruxelles, Service de Métrologie Nucléaire,50, av. F.D. Roosevelt, B-1050 Brussels, Belgium. Author's research is supported by the "Fonds National de la Recherche Scientifique" (Belgium) - Aspirant.

For $k = 0, 1 \ldots$ *until convergence do*

$$\begin{array}{llrcl}
(1.4) & \textit{Solve} & Bg_k & = & b - A\,x_k \\
(1.5) & & \delta_k & = & g_k + \sum\limits_{\substack{i=0\\k>i}}^{s-2} d_k^i \delta_{k-i-1} \\
(1.6) & & x_{k+1} & = & x_k + a_k\,\delta_k
\end{array}$$

where $s > 0$ is the order of the method ($\delta_k = g_k$ for first order schemes) and where the parameters a_k and d_k^i are determined by the choice of an actual method. It should be noticed that the system (1.4) to be solved at each step is consistent if and only if $\mathcal{R}(A) \subset \mathcal{R}(B)$, i.e. because symmetry, if and only if $\mathcal{N}(B) \subset \mathcal{N}(A)$, giving rise to the condition (1.3). This remark also shows that, in the singular case, there is no reason to enforce $\mathcal{N}(B) = \{0\}$, i.e. to consider only regular preconditionings. We shall actually see in the following that all classical iterative schemes used in the positive definite case do extend to the semidefinite case, with both singular and nonsingular preconditionings.

On the other hand, regarding preconditioning methods by modified incomplete factorizations, the algebraic framework of the regular theory, cfr [3], [4], [6], [7], [8], [9] , [10], [11], [12], [16], assumes that A is a nonsingular Stieltjes matrix (i.e. a positive definite matrix with nonpositive offdiagonal entries) and we shall show how this framework extends to the case of a singular Stieltjes matrix (i.e. a positive semidefinite matrix with nonpositive offdiagonal entries).

It will further follow from the general theory of polynomial acceleration that the spectral reduction technique (called spectral equivalence in the case of discrete PDE's) used in the regular case to reduce a general symmetric positive definite matrix A to Stieltjes one, also extends to the singular case.

General terminology and notation. All vectors belong to C^n; all matrices are n x n real matrices.

The symbol A^t, A^+, $\mathcal{N}(A)$, $\mathcal{R}(A)$, $\sigma(A)$ and $\rho(A)$ denote, respectively, the transpose, the Moore-Penrose inverse, the null space, the range, the spectrum and the spectral radius of the matrix A.

By $P_{M,L}$ we denote the projector with null space L and range M (this notation implying that L and M are complementary subspaces).

If A is an n x n matrix and T a subspace of C^n, we denote by $A\mid_T$ the linear operator in C^n defined as the restriction of A to T.

The order relation between real matrices and vectors is the usual componentwise order : if $A = (a_{ij})$ and $B = (b_{ij})$ then $A \leq B (A < B)$ if $a_{ij} \leq b_{ij} (a_{ij} < b_{ij})$ for all i, j; A is called nonnegative (positive) if $A \geq 0 (A > 0)$. If $A = (a_{ij})$, we denote by $diag(A)$ the (diagonal) matrix whose entries are $a_{ii}\delta_{ij}$ and we let $offdiag(A) = A - diag(A)$. By e we denote the vector with all components equal to unity; by a "(0,1) matrix", we understand a matrix whose nonzero entries are equal to unity.

We shall also need a few graph concepts; we refer to [15], [18] for general terminology about matrix graphs with the warning that, unless otherwise stated, all graphs considered here are ordered undirected graphs with node set [1,n], i.e. the ordered set of the first n integers or (when subgraphs are considered) some subset of [1,n]. In addition, we introduce the following more specific graph notions.

DEFINITION 1.1. *An increasing path in a graph is a path $i_o, i_1, i_2, ..., i_\ell$ such that $i_o < i_1 < i_2 < ... < i_\ell$.*

DEFINITION 1.2. *A node k of a graph G is clalled a precursor (successor) of the node i of G if $\{i, k\}$ belongs to the edge set of G with $k < i$ ($k > i$). The set of precursors (successors) of i is denoted by $P(i)$ ($S(i)$).*

DEFINITION 1.3. *For any node i of a graph G, we define the ascent $As(i)$ of i as*

$$As(i) = \{k \mid \text{There exists an increasing path from } k \text{ to } i\} .$$

It should be noticed that $i \in As(i)$ because a path of zero length is an increasing path.

DEFINITION 1.4. *For any pair of nodes i and j of a graph G, we denote by*

$$Pc(i,j) = P(i) \cap P(j)$$

their set of common precursors; we further define

$$Pc(G) = \bigcup_{\substack{i,j=1 \\ i \neq j}}^{n} Pc(i,j)$$

If G is the graph of an n x n matrix A, we also write $Pc(A)$ for $Pc(G)$.

2. Iterative schemes with possibly singular preconditionings.

2.1. Convergence analysis. For covering the singular case, we need to avoid assuming the matrices are invertible. The key of our ability to extend the regular theory lies thus in the use of generalized inverses. Some of their properties, needed for a good understanding of the paper, are recalled in the text. We refer to [13] for a more detailed exposition.

First, we assume in this paper that the operation (1.4) is achieved by the application of a linear operator, that is, there exists some matrix, say X, such that (1.4) reduces in practice to

$$g_k = X(b - Ax_k) \tag{2.1}$$

(as usual, it does not mean that X has to be computed explicitly). Now, $g = Xy$ is a solution of $Bg = y$ for all $y \in \mathcal{R}(B)$ if and only if $BXy = y$ for all $y \in \mathcal{R}(B)$, i.e. if and only if

$$B\,X\,B = B \tag{2.2}$$

that is, by definition, if and only if X is $\{1\}$-inverse of B. It follows then [13] that there exists some subspace S complementary to $\mathcal{N}(B)$ such that

$$X\,B = P_{S,\mathcal{N}(B)} \tag{2.3}$$

which means that for any $y \in \mathcal{R}(B)$, $g = Xy$ is the unique solution to $Bg = y$ that belongs to S. Finally, it turns out from these remarks that, for all $y \in \mathcal{R}(B)$

$$X\, y \;=\; P_{S,\mathcal{N}(B)} B^{(1)} y \tag{2.4}$$

where $B^{(1)}$ denotes any (other) $\{1\}$-inverse of B.

The Moore-Penrose inverse B^{+} of a matrix B is a particular $\{1\}$- inverse which verifies some additional properties. When B is symmetric, it is symmetric and can be characterized by its eigenvectors and eigenvalues which are such that

$$B^{+}x \;=\; \lambda^{+}\, x \;\Leftrightarrow\; Bx \;=\; \lambda\, x \tag{2.5}$$

with

$$\lambda^{+} \;=\; \begin{cases} \lambda^{-1} & \text{if } \lambda \neq 0 \\ 0 & \text{otherwise} \end{cases} \tag{2.6}$$

Using (2.4) with $B^{(1)} \;=\; B^{+}$, we can rewrite the iterative scheme (1.4-6)

$$g_k \;=\; P_{S,\mathcal{N}(B)}\, B^{+}(b - Ax_k) \tag{2.7}$$

$$\delta_k \;=\; g_k + \sum_{\substack{i=0 \\ k>i}}^{s-2} d_k^i\, \delta_{k-i-1} \tag{2.8}$$

$$x_{k+1} \;=\; x_k + a_k\, \delta_k \qquad\qquad k = 0, 1, 2... \tag{2.9}$$

Further, it is of interest to also include in our analysis the schemes in which the solution is at each step projected in a prescribed subspace T complementary to $\mathcal{N}(A)$, i.e. the schemes where the following is used rather than (2.7) :

$$g_k' = P_{S,\mathcal{N}(B)} B^{+}(b - Ax_k) \quad ; \qquad g_k = P_{T,\mathcal{N}(A)} g_k' \tag{2.10}$$

Finally, because (2.7), (2.10) are particular cases of

$$g_k = Q\, B^{+}(b - Ax_k) \tag{2.11}$$

where Q is some projector such that

$$\mathcal{N}(B) \subset \mathcal{N}(Q) \subset \mathcal{N}(A), \tag{2.12}$$

this formulation will be used from now on.

The scheme (2.11,8,9) is actually a polynomial acceleration of the basic unaccelerated iterative scheme

$$v_{k+1} \;=\; v_k + Q\, B^{+}(b - Av_k) \qquad k = 0, 1, 2... \tag{2.13}$$

and polynomially accelerated iterative schemes in the positive (semi)definite case with possibly singular preconditioning are studied in detail in [23]; we shall only briefly recall here the main issues of this work.

First, letting (x_k) be the sequence generated by (2.11,8,9) with given x_o, defining

$$(2.14) \qquad V = \mathcal{R}(B^+A) \qquad \text{and} \qquad \tilde{Q} = P_{V,\mathcal{N}(A)}$$

and letting $\tilde{x}_k = \tilde{Q}\, x_k$ for all k, it is stated that, for any (semi) norm $|\,.\,|$ whose kernel is $\mathcal{N}(A)$:

$$(2.15) \qquad | x_k - x | = |\tilde{x}_k - \tilde{x}|$$

for all x, $\tilde{x}$ such that $\tilde{x} = \tilde{Q}\, x$. Whence, the convergence analysis of the sequence (x_k) may equivalently be made on the sequence $(\tilde{x})$. It is further shown that the latter is identical to the sequence obtained when using the same iterative method (i.e. the same s, d_k^i, a_k) with starting vector $\tilde{x}_o = \tilde{Q}\, x_o$ in order to solve the regular system

$$(2.16) \qquad A\,|_V\; x = b$$

with the regular preconditioning

$$(2.17) \qquad (B^+\,|_{\mathcal{R}(A)})^{-1} = B\,|_V$$

With the additional result

$$(2.18) \qquad \sigma((B\,|_V)^{-1} A\,|_V) = \sigma(B^+A) \backslash \{0\},$$

we reach the conclusion that all classical methods used in the regular case for the determination of the iteration parameters d_k^i, a_k apply to our general iterative scheme (2.11,8,9), provided that "the extremal values of $\sigma(B^{-1}A)$" is understood as "the extremal values of $\sigma(B^+A)\backslash\{0\}$". By way of conclusion, we find of some interest to mention here that this reduction to the regular case is convenient but however not necessary for proving our results. Indeed, using generalized inverses in the classical proofs for the regular case, we can cover directly both the regular and singular cases. The interested reader may refer to [22].

2.2. Practical schemes. We have already mentioned that all classical methods apply to our more general feature without any change except in the way of defining the extremal eigenvalues associated with the preconditioning. We shall therefore not review them, referring the reader to the litterature [1], [3], [17], [20]. A brief summary in which the singular case is included is also presented in [23].

We however reach some additional results for the following methods.

The extrapolation method. It is a first ($s = 1$) order scheme. For such a scheme, letting

$$P_k(\nu) = \prod_{i=o}^{k-1} (1 - a_k \nu) \qquad \text{for} \quad k \geq 1$$

$$M_k = \max_{\nu \in \sigma(B^+A)\setminus\{0\}} | p_k(\nu) |$$

the error evolution formula is given by

$$| x - x_k |_A \leq M_k | x - x_o |_A$$

where $| \, . \, |_A = \sqrt{(.,A.)}$ denotes the A-(semi)norm and x any solution to (1.1). .The extrapolation method is associated with $a_k = \tau$ for all $k \geq 0$, the basic unaccelerated iterative method being included with $\tau = 1$. Letting ν_{min} and ν_{max} denote respectively the smallest and the largest value of $\sigma(B^+A)\setminus\{0\}$, the method is convergent if $\tau < 2/\nu_{max}$ with then $M_k = [max(| 1 - \tau\nu_{min} |, | \tau\nu_{max} - 1 |)]^k$. The optimum is thus reached for $\tau = 2/(\nu_{max} + \nu_{min})$, which leads to $M_k = [(\nu_{max} - \nu_{min})/(\nu_{max} + \nu_{min})]^k$.

When A is a regular Stieltjes matrix and $B = diag(A)$ (i.e. the point Jacobi preconditioner), the basic unaccelerated method ($\tau = 1$) is convergent (see [27]). Further, if the system to solve is the 5-point finite difference approximation of a second order elliptic PDE, $I - B^{-1}A$ is cyclic of index 2 [27] and therefore $\nu_{max} = 2 - \nu_{min}$, showing that the optimum is effectively reached for $\tau = 1$. On the other hand, when the PDE is a pure Neumann problem, A is (with the same 5-point finite difference discretization) a singular Stieltjes matrix (see definition below). With $B = diag(A)$, we have still that $I - B^{-1}A$ is cyclic of index 2, whence $\nu_{max} = 2$ and the unaccelerated method is no more convergent. If, however, one uses $a_o = 1/2$ with $a_k = 1$ for $k \geq 1$, then $p_k(\nu) = (1-\nu)^{k-1}(1-\nu/2)$, showing that this method is convergent if and only if $\nu_{max} \leq 2$ with $M_k = [max(| 1 - \nu_{min} |, | \nu'_{max} - 1 |)]^{k-1}$, where ν'_{max} denotes the maximal value of $\sigma(B^+A)\setminus\{2\}$. Moreover our assumptions imply $\nu'_{max} = 2 - \nu_{min}$, so that $M_k = [(\nu'_{max} - \nu_{min})/(\nu'_{max} + \nu_{min})]^{k-1}$, showing that the convergence rate is then about the same as in the regular case. The same remarks also hold for the block Jacobi preconditioner.

The steepest descent and the conjugate gradient method. These respectively first ($s = 1$) and second ($s = 2$) order methods present the particular feature that the coefficients a_k and d_k^o are computed during the iteration process according expressions which involve the vectors x_k and δ_k. Therefore, for that the convergence result given above applies, the coefficients are to be computed, in the singular case, not directly with the vectors x_k, δ_k, but with the vectors of the corresponding regular method which solve the regular system (2.16) with regular preconditioning (2.17). However, it is obtained in [23] that we may equivalently use the classical expressions with the vectors x_k, δ_k as they appear in the iterative scheme (2.11,8,9). Namely, for the conjugate gradient method, using

$$\text{(2.19)} \qquad a_k = \frac{(b - Ax_k, \delta_k)}{(\delta_k, A\delta_k)} \quad , \quad d_k^o = \frac{(b - Ax_k, g_k)}{(b - Ax_{k-1}, g_{k-1})} \quad ,$$

we obtain the coefficients which minimize the A-(semi)norm $| \, . \, |_A$ of the error among all the schemes (2.11,8,9). One can find in the litterature ([5],[26]) some developments about the convergence rate of this method in connection with the eigenvalue distribution. It turns out that all these developments extend to the singular case, provided that "the eigenvalue distribution in $\sigma(B^{-1}A)$" is understood as "the eigenvalue distribution in $\sigma(B^{+}A)\backslash\{0\}$". Note that (2.19) allows the same implementation techniques as in the regular case. A proof of the optimality properties that covers directly both the regular and singular cases can be found in [22].

2.3. Numerical stability. For studying stability, we have to consider, rather than (2.11,8,9)

$$
\begin{aligned}
&(2.20) \qquad & g_k &= QB^{+}(b - Ax_k) \\
&(2.21) \qquad & \delta_k &= g_k + \sum_{\substack{i=o \\ k>i}}^{s-2} d_k^i \delta_{k-i-1} \\
&(2.22) \qquad & x_{k+1} &= x_k + a_k\delta_k + \varepsilon_k \qquad k = 0, 1, 2, \ldots
\end{aligned}
$$

where ε_k, assumed to be small, also synthetises the perturbations that may appear in (2.20) or (2.21). Now defining the sequence $(\tilde{x}_k)$ by $\tilde{x}_k = \tilde{Q}\, x_k$ for all k (where $\tilde{Q}$ is defined by (2.14)), it follows from the above mentioned results of [23] that $(\tilde{x}_k)$ may be viewed as the sequence resulting from the perturbed iterative solution of the regular system $A\mid_V x = b$ with regular preconditioning $B\mid_V$. Thus, the stability analysis of an iterative method in the regular case applies to the sequence $(\tilde{x}_k)$. Further, one easily sees with Lemma 4.1 of [23] that for all $k \geq 1$

$$
(2.23) \qquad x_k = Q\,\tilde{x}_k + (I - Q\,\tilde{Q})x_o + \sum_{i=o}^{k-1} (I - Q\,\tilde{Q})\varepsilon_i
$$

showing that (2.20-22) (with ε_k assumed small) will be stable if and only if the corresponding regular method is. The stability analyses made for the regular case extend thus to the singular case. (One easily verifies in addition that the particular implementation we suggest above for the extrapolation method in the singular case is stable because $\mid 1 - a_k\nu \mid \leq 1$ for all $\nu \in \sigma(B^{+}A)$).

Besides these general aspects, Kaaschieter has pointed out in [19] that, in the singular case, some particular stability problems may appear when using the steepest descent or the conjugate gradient method. Indeed, we have to care that the condition $b \in \mathcal{R}(A)$ is not achieved in practice, due to roundoff errors. Generally, it is not cumbersome, because a small perturbation to b leads to a small term ε_k in (2.22), and if the iterative method used is stable, there will be no further problem. But, when using the steepest descent or the conjugate gradient method, we have to take into account that a_k is computed with the formula (2.19) in which b is involved. In order to see what may happen, let $b = b_o + \delta_b$, where $b_o \in \mathcal{R}(A)$ and $\delta_b \in \mathcal{N}(A)$, and let x be a solution to $Ax = b_o$. Then, for some given x_k and δ_k, we have

$$
\mid x_k + a_k\delta_k - x \mid_A^2 = \mid x_k - x \mid_A^2 + a_k^2 \mid \delta_k \mid_A^2 + 2a_k(\delta_k, A(x_k - x))
$$

so that we will have $\mid x_k + a_k\delta_k - x \mid_A \; \leq \; \mid x_k - x \mid_A$ if and only if

$$0 \le a_k \le \frac{2(\delta_k, A(x - x_k))}{(\delta_k, A\delta_k)} = \frac{2(\delta_k, b_o - Ax_k)}{(\delta_k, A\delta_k)}$$

or

$$0 \ge a_k \ge \frac{2(\delta_k, A(x - x_k))}{(\delta_k, A\delta_k)} = \frac{2(\delta_k, b_o - A_x)}{(\delta_k, A\delta_k)}$$

With $a_k = (\delta_k, b - Ax_k)/(\delta_k, A\delta_k)$ as in (2.19), we see that this condition will always be satisfied when $b = b_o$, while otherwise it is equivalent to

$$|\,(\delta_k, \delta_b)\,| \le |\,(\delta_k, b_o - Ax_k)\,| \tag{2.24}$$

When $b_o - Ax_k$ becomes small, (2.24) may be no more satisfied, so that the iterative process starts diverging, as observed in [19]. However, if δ_b is only due to the roundoff errors (as it is guaranteed by projecting the right hand side on the range of A), the stopping criterion will usually be met far before $\| b_o - Ax_k \| \approx \| \delta_b \|$, so that (2.24) will always be satisfied. On the other hand, if one wants to compute the solution with great accuracy, one can prevent problems by using (2.10) with $T = \mathcal{R}(A)$, so that $\delta_k \in \mathcal{R}(A)$ for all $k \ge 0$. Then, since $\delta_b \in \mathcal{N}(A)$, we will have $|\,(\delta_k, \delta_b)\,| = 0$ within the roundoffs errors, so that (2.24) will hold even when $\| b_o - Ax_k \| \approx \| \delta_b \|$.

2.4. Conclusion. It follows from the results of this section that all iterative methods effective in the regular case can be used in the singular case, with both regular and singular preconditionings. Moreover, in the case of singular preconditioning, the convergence properties of the scheme are completely independent of the choice of the generalized inverse, i.e. on the way chosen for achieving (1.4). The convergence results can be expressed as in the regular case, provided that "the eigenvalue distribution in $\sigma(B^+A)\backslash\{0\}$" is used as the right extension of "the eigenvalue distribution in $\sigma(B^{-1}A)$". This gives rise to the following generalized definition for the spectral condition number :

$$\kappa(B^+A) = \frac{\nu_{max}(B^+A)}{\nu_{min}(B^+A)} \tag{2.25}$$

with

$$\nu_{max}(B^+A) = \max_{\nu \in \sigma(B^+A)} \nu, \quad \nu_{min}(B^+A) = \max_{\substack{\nu \in \sigma(B^+A) \\ \nu \ne 0}} \nu \tag{2.26}$$

Finally, it is important to recall here the expressions obtained in [23] for ν_{max} and ν_{min}.

THEOREM 2.1. *Let A,B be symmetric nonnegative definite with $\mathcal{N}(B) \subset \mathcal{N}(A)$, and let $\nu_{max}(B^+A)$, $\nu_{min}(B^+A)$ be given by (2.26). We have*

$$\nu_{max} = \max_{\substack{z \in C^n \\ z \notin \mathcal{N}(B)}} \frac{(z, Az)}{(z, Bz)} \tag{2.27}$$

$$\nu_{min}(B^+A) = \min_{\substack{z\in\mathcal{R}(B^+A)\\ z\neq 0}} \frac{(z,Az)}{(z,Bz)}$$

$$(2.28) \qquad = \max_{\substack{S\\ S\oplus\mathcal{N}(A)=C^n}} \min_{\substack{z\in S\\ z\neq 0}} \frac{(z,Az)}{(z,Bz)}$$

Most results about spectral bounds are obtained, in the regular case, by proving inequalities of the type $(z,Az) \leq c_1(z,Bz)$ or $(z,Az) \geq c_2(\lambda_1)(z,Bz)$ for all $z \in C^n$, where λ_1 is the first eigenvalue of $D^{-1}A$, with $D = diag(A)$; see [3] for examples. It follows from Theorem 2.1 that the first type of inequality will give also an upper bound in the singular case, while, as will be seen below, the inequalities of the second type can generally be rewritten $(z,Az) \geq c_2(\lambda_{min})(z,Bz)$, where λ_{min} is the first nonzero eigenvalue of $D^{-1}A$, so that they give also a lower bound in the singular case.

Therefore, many regular conditioning analysis results can be rewritten with little handling so as to cover both the regular and singular cases. However, regarding the modified incomplete factorization methods, the generalization is not straightforward because the classical existence theorems do not work in the singular case. We summarize in the next section our new existence analysis that allows covering the singular case and, in Section 4, our results on conditioning analysis and factorization strategies.

3. Modified incomplete factorizations of Stieltjes matrices. In the regular case, the basic framework of the modified incomplete factorization methods assumes that the system matrix is a (regular) Stieltjes matrix. We first need to appropriately extend this notion to the singular case. To this aim, we use the following definition.

DEFINITION 3.1. *A real square matrix A is called an M-matrix if there exists a non negative number t such that*

$$tI - A \geq 0 \qquad \textit{with} \qquad \rho(tI-A) \leq t;$$

a symmetric M-matrix is called a Stieltjes matrix.

It follows from Definition 3.1 that a Stieltjes matrix has nonpositive offdiagonal entries and that a symmetric matrix with non positive offidiagonal entries is a Stieltjes matrix if and only if it is nonnegative definite. The latter property has already been assumed in Section 1. The restriction concerning the offdiagonal entries is thus the only one we have to introduce from now on. However, it should be mentioned here that the modified incomplete factorization methods have actually a much wider scope of application. Indeed, the spectral reduction technique used in the regular case, called spectral equivalence in discrete PDE's applications ([3], [16]), readily extends to the semidefinite case. With this technique, an arbitrary symmetric positive (semi)definite matrix A may be reduced to the Stieltjes case provided that one can determine a Stieltjes matrix A_o and positive numbers α, β with

$$(3.1) \qquad \alpha(z, A_oz) \leq (z, Az) \leq \beta(z, A_oz)$$

for all $z \in C^n$; then, we have obviously $\mathcal{N}(A) = \mathcal{N}(A_o)$ and

$$\kappa(B^+A) \leq \frac{\beta}{\alpha}\,\kappa(B^+A_o)$$

readily follows from Theorem 2.1 for any preconditioner B such that $\mathcal{N}(B) \subset \mathcal{N}(A_o)$. Note that in [3], [16], inequalities of the type (3.1) are derived for finite element matrices by analysing element stiffness matrices. The latter are singular so that such analyses do apply to the singular case without any modification. The Stieltjes theory allows by this way to cover most discrete PDE's applications. Its extension to the singular case means, with this respect, that the pure Neumann problems are no more excluded.

The modified incomplete factorizations of a Stieltjes matrix A are based, in the regular case, upon the existence for such matrices of a positive vector x such that $Ax \geq 0$. The following theorem extends this result to the singular case (see [14] for a proof).

THEOREM 3.2. *Let $A = (a_{ij})$ be a Stieltjes matrix. Then, there exists a positive vector x such that $Ax \geq 0$. Further*
(1) *If A is regular : $\exists x > 0 : \sum_{j=1}^{i} a_{ij}x_j > 0$ for all i*
(2) *If A is irreducible and singular :*
  a) $\exists x > 0 : \mathcal{N}(A) = Span\{x\}$
  b) $\forall x : Ax \geq 0 \Rightarrow Ax = 0$

We recall now from [24] our definition of modified incomplete factorizations.

DEFINITION 3.3. *Let $A = (a_{ij})$ be a $n \times n$ Stieltjes matrix and let $x > 0$ be such that $Ax \geq 0$; let $\Lambda = (\lambda_i\delta_{ij})$ be a nonnegative diagonal matrix and $\beta = (\beta_{ij})$ a (0,1) matrix; let $U = (u_{ij})$ be the $n \times n$ upper triangular matrix defined by the following algorithm : for i=1,...,n set*

$$u_{ij} = a_{ij} - \beta_{ij}\sum_{k<i} u_{ki}u^+_{kk}u_{kj} \qquad\qquad for\ i < j \leq n$$

$$\begin{cases} (Ux)_i & = (Ax)_i + \lambda_i a_{ii}x_i - \sum_{k<i} u_{ki}u^+_{kk}(Ux)_k \\ u_{ii} & = ((Ux)_i - \sum_{j>i} u_{ij}x_j)/x_i \end{cases}$$

where u^+_{kk} is defined by

$$u^+_{kk} = \begin{cases} u_{kk}^{-1} & if \qquad u_{kk} \neq 0 \\ 0 & if \qquad u_{kk} = 0 \end{cases}$$

We say that U is the upper triangular factor and $P = diag(U)$ the diagonal factor of the modified incomplete factorization

$$B = U^tP^+U$$

of A associated with x, Λ and β.

Note that by (2.5), (2.6) we have $P^+ = (u^+_{ii}\delta_{ij})$. Λ is often referred as the perturbation term, the factorizations with $\Lambda = 0$ being called unperturbed factorizations.

The following theorem contains our existence analysis.

THEOREM 3.4. *Let A be an irreducible $n \times n$ Stieltjes matrix and x a positive vector such that $Ax \geq 0$. Let U be the upper triangular factor of the modified incomplete factorization $B = U^tP^+U$ (with $P = diag(U)$) of A associated with x, Λ and β,* *where* Λ *is a nonnegative diagonal matrix and* β *a (0,1) matrix, and let* $D = diag(A)$; *then :*

(1) *U is an upper triangular M-matrix with* $Ux \geq Ax \geq 0$;

(2) *B is symmetric and nonnegative definite with* $Bx = Ax + \Lambda Dx$;

(3) $\mathcal{N}(B) = \mathcal{N}(U)$;

(4) $\mathcal{N}(B) \subset \mathcal{N}(A)$ *if and only if, in the graph of U :*

$$\begin{cases} \forall i < n : S(i) \neq \phi \\ \underline{or} \\ \forall i : S(i) = \phi \;\Rightarrow\; \exists j \in As(i) : (Ax + \Lambda Dx)_j > 0; \end{cases}$$

(5) *if* $\mathcal{N}(B) \subset \mathcal{N}(A)$ *:*

$$\mathcal{N}(B) = \begin{cases} \mathcal{N}(A) & \text{if} \quad \Lambda = 0 \\ \{0\} & \text{otherwise.} \end{cases}$$

Proof. Statements (1), (2), (3), (5) are proven in [24]. The first sufficient condition (4) is also proven in [24] while the second readily implies by induction that $u_{ii} > 0$ for all i, hence $\mathcal{N}(B) = \{0\}$. For proving the necessity of condition (4), note first that if the second criterion is not met, there exists some i with $u_{ii} = 0$, hence B is singular by (3). But by (5), we have that B singular is compatible with $\mathcal{N}(B) \subset \mathcal{N}(A)$ if and only if A is singular and $\Lambda = 0$, and in the latter case, the necessity of the condition $i < n \Rightarrow S(i) \neq \phi$ is proven in [24]. ∎

Note that the classical existence criterion used generally in the regular case to ensure $\mathcal{N}(B) = \{0\}$ (that is the semi strict diagonal dominance criterion [11]) assumes

$$\sum_{j=1}^{i} a_{ij} x_j > 0,$$

which implies

$$\forall i \;:\; S(i) = \phi \Rightarrow (Ax)_i > 0$$

and appears therefore clearly to be included in the second condition of (4) since $i \in As(i)$ by definition. The same remark also holds for Gustafsson existence analysis which only considers the cases where $\Lambda Dx > 0$. However, both do not cover the case of A singular with $\Lambda = 0$, giving rise to our first criterion which is also useful in the regular case because there are no more restrictions on the entries of A nor on the perturbation Λ, all being replaced by a graph condition to be satisfied by U. Since the graph of U includes that of A, it is fulfilled by U when it is by A; otherwise, it requires either some fill-in of the matrix U or an appropriate reordering of the system matrix; note with this respect that the ordering procedure described in [10] leads anyway to a system matrix satisfying $S(i) \neq \phi$ for all $i < n$. Finally, an alternate technique, introduced in [24], may be used to prove that $\mathcal{N}(B) \subset \mathcal{N}(A)$ which simply consists in checking

$$(z, Az) \leq c(z, Bz)$$

for all $z \in C^n$. The existence analysis may then be viewed as a part of the conditioning analysis, the existence being guaranteed for any factorization algorithm or "strategy" for which an upper eigenvalue bound is obtained.

Definition 3.3 leaves open the choice of x, Λ and β. The choice of x is limited in the regular case by practical considerations, while there is no choice at all in the singular case. About β, its optimization would require to allow the fill-in of U up to the point where the reduction of the number of iterations is compensated by the increase of the algorithm computational complexity. The choice of β may also be limited by the memory requirements and the data structure one gives to the computer program. In practice, one uses either the "incomplete factorization by position", for which β is *a priori* fixed, or the "incomplete factorization by value" in which β is dynamically determined. Such aspects lie outside the scope of the present work, and we shall concentrate in the next section on the choice of Λ, assuming that x and β are given. It should be noted that this assumption is actually not incompatible with a dynamic determination of β, because the latter deals with the offdiagonal entries of the approximate factor U, while Λ plays only a role in the determination of its diagonal entries.

4. Factorization strategies and conditioning analysis. We extend here to the singular case the factorization algorithms (or "strategies") developped in [10]. We shall state their conditioning analysis together with their other properties referring to [21], [24], [25] for a detailed exposition of eigenvalue bounds which includes the singular case. It should be noticed that, with the results of these works and those of the preceding sections, most remarks and comments made in [10] for the regular case actually apply to our more general framework. We shall therefore give a somewhat abrupt exposition, stressing only on the particularities involved by the singular case. On the other hand, it clearly follows from [10] that the strategies No.2, 3 and 4 proposed there are not essentially different. For brevity, we shall therefore only generalize here the strategy No.4 (our strategy No.2).

In each case, $A = (a_{ij})$ is a Stieltjes matrix, x a given positive vector such that $Ax \geq 0$ and $\beta = (\beta_{ij})$ a given (0,1) matrix.

Strategy No.1. It consists in choosing $\Lambda = 0$, i.e. that B is the modified incomplete factorization of A associated with $x, \Lambda = 0$ and β. (See Definition 3.3 for the factorization algorithm). This strategy is also referred to as the unperturbed strategy. Its properties are displayed in the following theorem (see [24] for a proof).

THEOREM 4.1. *Let A be an irreducible Stieltjes matrix and U the upper triangular factor of the modified incomplete factorization $B = U^t P^+ U$ (with P=diag(U)) of A associated with x, Λ and β, where x is a positive vector such that $Ax \geq 0, \beta$ a (0,1) matrix and $\Lambda = 0$. Let $0 \leq \tau \leq 1$ be given by*

$$\tau = \begin{cases} \max_{i \in Pc(U)} & \frac{((P-U)x)_i}{(Px)_i} & \textit{if } Pc(U) \neq \phi \\ 0 & & \textit{otherwise} \end{cases}$$

We have :

(1) *If $\tau < 1$,*

(4.1) $$\mathcal{N}(B) \subset \mathcal{N}(A)$$

with

(4.2) $$\nu_{max}(B^+ A) \leq \frac{1}{1-\tau}$$

(2) *If $\mathcal{N}(B) \subset \mathcal{N}(A)$*

(4.3) $$\mathcal{N}(B) = \mathcal{N}(A)$$

with

(4.4) $$\nu_{min}(B^+A) \geq 1$$

(3) *In particular, if A is singular :*

a) $Bx = Ux = Ax = 0$

b) $\mathcal{N}(B) \subset \mathcal{N}(A)$ *if and only if* $i < n \Rightarrow S(i) \neq \phi$

Depending on the case at hand, the bound (4.2) may be satisfying or not (see [10]). When A is singular, it however clearly follows from (3) that $\tau = 1$ except in the trivial case where $Pc(U) = \phi$ and therefore $B = A$. The bound (4.2) is thus useless in the singular case. Now, the conditioning analysis of this strategy has been improved in [9], [10], giving an upper bound compatible with $\tau = 1$, the diagonal dominance requirement being replaced by a graph condition to be satisfied by the graph of U. This theory has been further improved and extended to the singular case in [25]. It is not possible to summarize briefly these results because they require the introduction of a somewhat involved formalism; we therefore refer the reader to the above quoted works.

On the other hand, it follows from Theorem 4.1 that this strategy leads in the singular case to a singular preconditioning. As shown in Section 2, this is not anyway disturbing providing that we can achieve the step (1.4) : "solve $Bg = y$" for any $y \in \mathcal{R}(B)$ by the use of a linear operator. Now, letting $\tilde{U}$ be obtained from U by "shifting" (in the singular case) its last diagonal entry up to an arbitrary positive number, and setting $\tilde{P} = diag(\tilde{U})$, it can be shown that when $\mathcal{N}(U) = Span\{x\}$, $\tilde{U}, \tilde{P}$ and $\tilde{B} = \tilde{U}^t \tilde{P}^{-1} \tilde{U}$ are regular matrices such that $g = \tilde{B}^{-1} y$ provides a solution to $Bg = y$ for all $y \in \mathcal{R}(B)$. In other words $\tilde{B}^{-1}$ is a $\{1\}$-inverse of B (cfr. [24] for a proof). From a practical point of view, it should be stressed that this "shift" is the only modification to implement in a computer program (that uses the PCG algorithm with the unperturbed modified incomplete factorization) to get it working for singular problems too.

Now, even with the above mentioned improved conditioning analysis, the upper bound obtained for this strategy may still be unsatisfying, giving rise to the following strategy, in which an upper bound is anyway guaranted, at the expense of a decrease of the smallest eigenvalue $\nu_{min}(B^+A)$.

Strategy No.2. It consists in computing the upper triangular matrix $U = (u_{ij})$ according to the following algorithm, where $0 < \tau < 1$ is an *a priori* chosen parameter.

ALGORITHM 1.

For $i = 1,\dots,n$ set :

$$u_{ij} = a_{ij} - \beta_{ij} \sum_{k \in Pc(i,j)} \frac{u_{ki}u_{kj}}{u_{kk}}, \quad j = i+1, \dots, n$$

$$u_{ii}x_i = \begin{cases} \max(\tau^{-1} \sum_{j>i} u_{ij}x_j, (Ax)_i - \sum_{j>i} u_{ij}x_j \\ \qquad - \sum_{k \in P(i)} u_{ki}u_{kk}^{-1}(Ux)_k) & \text{if } i \in Pc(U) \\ (Ax)_i - \sum_{j>i} u_{ij}x_j - \sum_{k \in P(i)} u_{ki}u_{kk}^{-1}(Ux)_k & \text{if } i \notin Pc(U) \end{cases}$$

It turns out that this algorithm can never break down since we will have $u_{ii} > 0$ for all i such that $u_{ij} \neq 0$ for some $j > i$.

The preconditioner is then given by $B = U^t P^+ U$, where $P = diag(U)$. Its properties are stated in the following theorem (see [24] for a proof).

THEOREM 4.2. *Let A be an irreducible Stieltjes matrix, x a positive vector such that $Ax \geq 0$ and β a (0,1) matrix. Let U be the upper triangular matrix computed according Algorithm 1, where $0 < \tau < 1$ is some given parameter. Set $P = diag(U)$ and $B = U^t P^+ U$. We have :*

(1) *There exists a nonnegative diagonal matrix $\Lambda = (\lambda_i \delta_{ij})$ such that U is the upper triangular factor of the modified incomplete factorization B of A, with respect to x, Λ and β. Further,*

$$\lambda_i a_{ii} x_i \begin{cases} = 0 & \text{if } i \notin Pc(U) \\ \leq max(0, \tau^{-1}(1-\tau)^2((P-U)x)_i + \\ \quad (1-\tau)((U^t - U)x)_i - (Ax)_i) & \text{if } i \in Pc(U) \text{ with } P(i) \subset Pc(U) \\ \leq max(0, \tau^{-1}((P-U)x)_i - (Ax)_i) & \text{otherwise} \end{cases}$$

(2) *Except in the trivial case where $Pc(U) = \phi$ and therefore $B = A, U, P$ and B are regular with*

$$\nu_{max}(B^+ A) \leq \frac{1}{1-\tau} \tag{4.5}$$

and

$$\nu_{min}(B^+ A) \geq \Big(1 + \min_{\substack{S \\ S \oplus \mathcal{N}(A) = C^n}} \max_{\substack{z \in S \\ z \neq 0}} \frac{(z, \Lambda D z)}{(z, Az)}\Big)^{-1} \tag{4.6}$$

where $D = diag(A)$.

Note that the factorization algorithm proposed by Axelsson-Barker in [3] (eq. 7.18) is also based upon the choice of a parameter $\alpha > 0$ such that $1/\alpha$ is an upper bound on $\nu_{max}(B^+ A)$ Clearly, the efficiency of such strategies depends on the smallest eigenvalue ($\nu_{min}(B^+ A)$) behaviour. Now, one easily verifies with (1) of Theorem 4.2 that the perturbations λ_i involved by Algorithm 1 are anyway smaller [1] than those required for achieving Axelsson-Barker scheme when one uses $\alpha = 1 - \tau$ in the latter (which leads to the same value for the upper bound). Therefore, the smallest eigenvalue analysis developped in [3] also applies to Strategy No.2. It consists essentially in the proof of an inequality of the type $(z, \Lambda D z) \leq c_1(z, Dz) + c_2(z, Az)$ which, extended to the singular case, leads together with (4.6) to $\nu_{min}(B^+ A) \geq 1/(1 + c_2 + c_1 \lambda_{min}^{-1})$ where λ_{min} denotes the first nonzero eigenvalue of $D^{-1}A$.

This extension is included in an improved version of the Axelsson-Barker result, proven in [24]. Our improvements allow first to remove some hypotheses of the original version and secondly, by a more accurate analysis, to obtain bounds that are also good approximations of ν_{min} (see [21]). The major conclusion we can deduce from these technical results is that, applied to discrete second order elliptic PDE's, for a family

[1] except possibly at nodes $i \in Pc(U)$ such that $Pc(U)$ does not contain $P(i)$; such nodes are however usually not met in practice

of problems $A_h x_h = b_h$ which differ only by the mesh size h, chosing $\tau_h = 1 - c_1 h$, they give $\nu_{min}(B_h^+ A_h) \geq c_2$, where c_2 is independent of h; further, with (4.5), this leads to $\kappa(B_h^+ A_h) \leq 0(h^{-1})$.

To avoid the difficulties that would be involved by the use of the above mentioned results and other similar [8], [21] for a practical estimation of ν_{min}, it is proposed in [10] an heuristic estimate. According to our general philosophy, we propose to generalize the latter by simply replacing $\lambda_1(D^{-1}A)$ by $\lambda_{min}(D^{-1}A)$; i.e. we propose

$$\nu_{min} \cong \frac{1}{1 + \frac{<\lambda>}{\lambda_{min}}} \tag{4.7}$$

where

$$< \lambda > = \frac{(x, \Lambda D x)}{(x, D x)} \tag{4.8}$$

and

$$\lambda_{min} = \min_{\substack{\lambda \in (D^{-1}A) \\ \lambda \neq 0}} \lambda \tag{4.9}$$

As in the regular case, (4.7) is justified from Theorem 4.2 by stating that, letting z_{min} be such that $Az_{min} = \lambda_{min} z_{min}$,

$$\min_{\substack{S \\ S \oplus \mathcal{N}(A) = C^n}} \quad \max_{\substack{z \in S \\ z \neq 0}} \frac{(z, \Lambda D z)}{(z, A z)} \cong \frac{(z_{min}, \Lambda D z_{min})}{(z_{min}, A z_{min})} \tag{4.10}$$

and

$$\frac{(z_{min}, \Lambda D z_{min})}{(z_{min}, D z_{min})} \cong \frac{(x, \Lambda D x)}{(x, A x)} \tag{4.11}$$

generally hold for the matrices Λ as they appear in practical applications of approximate factorization algorithms like Algorithm 1.

For this strategy to be useful in practice, we have still to provide some rules for the determination of τ. To this aim, let use the ratio of the bound (4.5) by the estimate (4.8) as estimate of $\kappa(B^+ A)$, and let further use .

$$(x, \Lambda D x) \cong (x, \frac{(1-\tau)^2}{\tau}(P - U)x + (1 - \tau)[max(0, U^t - U)]x), \tag{4.12}$$

$$(x, (P - U)x) \cong \frac{1}{2}(x, Dx); \tag{4.13}$$

we have then, with $\alpha = (1 - \tau)$

$$\kappa(B^+A) \cong \frac{1}{\alpha} + \frac{\alpha}{1-\alpha}\frac{1}{2\lambda_{min}} + \gamma \tag{4.14}$$

where γ is some constant independent of τ. The optimum is reached in (4.14) when

$$\alpha^2 = 2\lambda_{min}\frac{1-2\alpha}{(1-\alpha)^2},$$

i.e. when $\lambda_{min} << 1$:

$$1-\tau = \alpha \cong \sqrt{2\lambda_{min}} \tag{4.15}$$

Then, (4.14) becomes

$$\kappa(B^+A) \cong \sqrt{2\lambda_{min}} + \gamma \cong \kappa(D^{-1}A) + \gamma \tag{4.16}$$

showing that Strategy No.2 allows generally when γ is not too large, to improve the original conditioning by an order of magnitude. Unfortunately, λ_{min} is generally not known in advance so that (4.15) can generally not be achieved in practice. In discrete PDE's applications however, we have $\lambda_{min} = 0(h^{-2})$ so that we can deduce from (4.15) the following rule

$$1-\tau = \xi\frac{hS}{4V} \tag{4.17}$$

where h denotes the - or a typical - mesh size, V the area (volume) of the domain and S the length (area) of its boundary, and where ξ is a parameter. Our present experiments indicate that ξ has to be chosen not far from unity, $\kappa(B^+A)$ being in that case generally relatively insensitive to its actual value. When using the preconditioned conjugate gradient process, we further observe that a sharp optimization of $\kappa(B^+A)$ is not needed, due to the optimal convergence properties of the method. This is illustrated in the following section. Note that (4.17) is compatible with the above mentioned results $\nu_{min} \geq 0(1)$, $\nu_{max} \leq 0(h^{-1})$ and $\kappa(B^+A) \leq 0(h^{-1})$.

5. Example. We apply here the results of the preceding sections to the finite difference approximation of the Neumann problems

$$-\bar{\nabla}\, D(x,y)\, \bar{\nabla}\, u = f \quad \text{on } \Omega =]0,1[\times]0,1[$$
$$D\frac{\partial u}{\partial n} = 0 \qquad \text{on } \partial\Omega$$

where $D(x,y)$, $0 \leq x,y \leq 1$, is given by

Problem 1 :

$$D(x,y) = 1 \text{ for all } 0 \leq x,y \leq 1$$

Problem 2 :

$$D = \begin{cases} .010 & \text{if } 0 \leq x,y < \frac{1}{3} \\ 1000 & \text{if } \frac{2}{3} < x,y \leq 1 \\ 1 & \text{otherwise} \end{cases}$$

Problem 3 :

$$D = \begin{cases} 100 & \text{if } 0 \le x, y < \frac{1}{3} \\ 1000 & \text{if } \frac{2}{3} < x, y \le 1 \\ 1 & \text{otherwise} \end{cases}$$

In each case, we use an uniform mesh size $h = 1/N$, the order of the resulting linear system $Ax = b$ being $n = (N+1)^2$. A is ordered according to the lexicographic ordering and we consider only modified incomplete factorizations with $\beta = 0$, while the positive vector to be used in e because $\mathcal{N}(A) = Span\{e\}$. For both strategies proposed in section 4 and for the Incomplete Cholesky preconditioner, we have computed ν_{max} and ν_{min} (given by (2.26)) for N = 12, 24, 48, 96. We also report the number k of preconditioned conjugate gradient (PCG) iterations necessary for meeting the relative residual error $\|r_k\|/\|r_o\| \le \varepsilon$, for $\varepsilon = 10^{-3}, 10^{-5}, 10^{-8}$. In each case, $b = Au$, where u is the vector that samples the function $(1+x)^2(1+y)(2-y)e^{xy}$. The initial approximation used is $x_o = 0$.

Strategy No.1. $U = (u_{ij})$ is determined by

$$offdiag(U + U^t) = offdiag(A) \ \ and \ \ Ue = 0$$

and we set $P = diag(U)$, $B = U^t P^+ U$. We have $\mathcal{N}(B) = \mathcal{N}(A)$, and, in order to perform iterations, we have still to choose a $\{1\}$- inverse of B. For this purpose, we define $\widetilde{U} = (\widetilde{u_{ij}})$ by $\widetilde{u_{ij}} = u_{ij}$ for all i, j, except $i = j = n$ while $\widetilde{u_{nn}} = 1$ and we set $\widetilde{P} = diag(\widetilde{U})$ and $\widetilde{B} = \widetilde{U}^t \widetilde{P}^{-1} \widetilde{U}$. As mentioned in Section 4, $\widetilde{B}^{-1}$ is then a $\{1\}$-inverse of B and we use it for the solution of our test problem by the PCG process. For completeness, we also mention the upper bound that would be obtained for this example by applying the results of [25].

Strategy No.2. U is here computed according to Algorithm 1, where τ is determined by (4.17) (i.e. $\tau = 1 - \xi/N$) with $\xi = .5, 1., 2.$. We also mention the estimate obtained for ν_{min} by the use of (4.7).

Incomplete Cholesky. Kaasschieter has proposed in [19] the use of the Incomplete Cholesky (IC) preconditioner for solving singular systems. For comparison purpose, we have included the latter in our experiments. Actually, one easily verifies (see [4]) that the Incomplete Cholesky preconditioner is identical to that obtained by a modified incomplete factorization with $x = e$, $\beta = 0$ and $\Lambda = (\lambda_i \delta_{ij})$ (dynamically) determined by

$$\lambda_i a_{ii} = \sum_{k \in P(i)} \sum_{\substack{j \in S(k) \\ j \ne i}} \frac{a_{ki} a_{kj}}{u_{kk}}. \tag{5.1}$$

This allows us to use (4.7) to estimate ν_{min}.

TABLE 1
Results for Problem 1

		ν_{min}	est.	ν_{max}	up.b.	$\kappa(B^+A)$	Num. of iterations 10^{-3}	10^{-5}	10^{-8}
	S1	1.	1.	32.	48.	32.	12	17	25
N=12	ξ=.5	.83	.87	8.2	24.	9.9	10	15	21
n=169	ξ=1.	.65	.71	5.2	12.	8.0	10	15	21
	ξ=2.	.38	.43	3.0	6.	7.8	10	15	21
	IC	.11	.10	1.2		11.5	11	16	22
	S1	1.	1.	70.	96.	70.	17	27	39
N=24	ξ=.5	.83	.87	17.	48.	20.	15	22	32
n=625	ξ=1.	.65	.71	10.	24.	16.	14	20	29
	ξ=2.	.40	.44	5.8	12.	15.	14	21	29
	IC	.28E-1	.28E-1	1.2		43.	19	30	38
	S1	1.	1.	150.	192.	150.	26	40	62
N=48	ξ=.5	.83	.87	34.	96.	41.	21	32	47
n=2401	ξ=1.	.66	.71	21.	48.	32.	20	29	42
	ξ=2.	.40	.44	11.	24.	29.	19	29	40
	IC	.73E-2	.73E-2	1.2		168.	36	55	70
	S1	1.	1.	242.	384.	242.	41	63	97
N=96	ξ=.5	.83	.87	70.	192.	84.	30	47	70
n=9409	ξ=1.	.66	.71	42.	96.	64.	29	42	61
	ξ=2.	.40	.44	23.	48.	57.	27	41	58
	IC	.18E-2	.18E-2	1.2		668.	71	95	136

The results are displayed in Tables 1,2,3 (S1 refers to Strategy No.1, ξ=.5, 1., 2. to Strategy No.2 with the corresponding value of ξ; est. refers to the estimate (4.7) of ν_{min}; up.b. refers to the upper bound obtained in [25] when the Strategy No.1 is concerned, and to the upper bound (4.5) when it is Strategy No.2). The following remarks turn out :

- For the Strategy No.2, we observe that the value of ξ in (4.17) has little influence on $\kappa(B^+A)$ (except for Problem 3) and nearly not on the number of iterations.

- For Problems 1 and 2, both Strategies No.1 and 2 present a spectral condition number $0(h^{-1})$, entailing a number of iterations bounded by $0(h^{-1/2})$. The Strategy No.2 gives better results for Problem 1, while, for Problem 2, in spite of an higher spectral conditioning, the Strategy No.1 gives somewhat better results for realistic stopping criterions. For small ε, however, the Strategy No.2 is still better. Such behaviours can be explained by the superlinear convergence of the PCG process in connection with the eigenvalue distribution in $\sigma(B^+A)\setminus\{0\}$ (see [4], [5], [26]).

TABLE 2
Results for Problem 2

		ν_{min}	est.	ν_{max}	up.b.	$\kappa(B^+A)$	Num. of iterations 10^{-3}	10^{-5}	10^{-8}
	S1	1.	1.	14.	48.	14.	7	12	18
N=12	ξ=.5	.60	.55	5.7	24.	9.5	8	13	20
n=169	ξ=1.	.37	.34	4.1	12.	11.	9	13	20
	ξ=2.	.17	.16	2.7	6.	15.7	9	14	21
	IC	.42E-1	.42E-1	1.2		29.5	11	15	21
	S1	1.	1.	33.	96.	33.	11	18	29
N=24	ξ=.5	.60	.53	11.	48.	19.	12	20	28
n=625	ξ=1.	.38	.33	7.9	24.	21.	13	19	29
	ξ=2.	.18	.16	5.0	12.	29.	12	20	30
	IC	.11E-1	.11E-1	1.2		114.	21	29	40
	S1	1.	1.	74.	192.	74.	17	27	46
N=48	ξ=.5	.60	.51	24.	96.	39.	18	29	41
n=2401	ξ=1.	.38	.32	16.	48.	44.	18	27	40
	ξ=2.	.18	.16	10.	24.	56.	18	28	42
	IC	.27E-2	.27E-2	1.2		455.	39	56	75
	S1	1.	1.	165.	384.	165.	25	42	69
N=96	ξ=.5	.60	.51	51.	192.	84.	27	40	61
n=9409	ξ=1.	.38	.32	34.	96.	91.	27	38	59
	ξ=2.	.18	.15	21.	48.	114.	25	35	59
	IC	.70E-3	.70E-3	1.2		1819.	76	113	148

- The Problem 3 presents some pathological features. First, the upper bound obtained in [25] for the unperturbed Strategy is no more $0(h^{-1})$, and the actual value of ν_{max} as well. Therefore, in spite of a nice value for κ obtained for moderate h, the method is less interesting with respect to very small h. Second, the first non zero eigenvalue of $D^{-1}A$ (with $D = diag(A)$) is about 100 times less than those of Problem 1 (it is a "quasi-singular" singular problem). This explains, by the relation (4.7), the very small values obtained for ν_{min} with the Strategy No.2. In spite of its behaviour which is still $0(h^{-1})$, $\kappa(B^+A)$ presents then a very high value. However, comparing the convergence behaviour with those observed for Problems 1 and 2, we see that only a very few extra iterations are needed. Again, this has to be explained by the super linear convergence in connection with the eigenvalue distribution. We further observe that this super linear convergence is particularly effective for small h and ε, i.e. when the Strategy No.1 fails to be completely satisfying.

- Regarding to the "model" Problem 1, we see that the number of iterations involved by the IC preconditioner is prohibitive for small and even moderate h. This is due to the spectral conditioning which is $0(h^{-2})$ for this method. This behaviour is readily explained by our estimate (4.7) (which is found accurate enough). Indeed, by (5.1) we have $< \lambda >= 0(1)$, so that (4.7) gives, with $\lambda_{min} << 1$:

TABLE 3
Results for Problem 3

		ν_{min}	est.	ν_{max}	up.b.	$\kappa(B^+A)$	Num. of iterations 10^{-3}	10^{-5}	10^{-8}
	S1	1.	1.	17.	100.	17.	8	13	19
N=12	ξ=.5	.38E-1	.38E-1	5.7	24.	152.	13	16	24
n=169	ξ=1.	.16E-1	.16E-1	4.1	12.	251.	12	16	22
	ξ=2.	.63E-2	.63E-2	2.7	6.	428.	13	17	23
	IC	.14E-2	.14E-2	1.2		877.	15	18	23
	S1	1.	1.	41.	311.	41.	13	20	33
N=24	ξ=.5	.35E-1	.35E-1	11.	48.	315.	18	24	33
n=625	ξ=1.	.15E-1	.15E-1	7.8	24.	508.	17	23	32
	ξ=2.	.61E-2	.61E-2	5.0	12.	831.	18	24	32
	IC	.35E-3	.35E-3	1.2		3497.	28	33	42
	S1	1.	1.	110.	934.	110.	19	33	53
N=48	ξ=.5	.34E-1	.34E-1	23.	96.	692.	27	35	49
n=2401	ξ=1.	.15E-1	.15E-1	16.	48.	1090.	24	33	45
	ξ=2.	.59E-2	.59E-2	10.	24.	1701.	25	34	45
	IC	.87E-4	.87E-4	1.2		14030.	54	64	80
	S1	1.	1.	334.	2690.	334.	32	53	86
N=96	ξ=.5	.33E-1	.33E-1	50.	192.	1509.	38	50	71
n=9409	ξ=1.	.15E-1	.15E-1	34.	96.	2311.	36	47	67
	ξ=2.	.59E-2	.59E-2	21.	48.	3498.	37	49	63
	IC	.22E-4	.22E-4	1.2		56198.	105	127	156

$$\nu_{min}(B^+A) \cong \lambda_{min}/ <\lambda>= 0(\lambda_{min})$$

showing that the IC preconditioner fails to improve the original conditioning of an order of magnitude as both Strategies No.1 and 2 do. We further observe that, when passing from the "model" Problem 1 to the more realistic Problems 2 and 3, the convergence behaviour associated with the IC preconditioner deteriorates much more than those associated with the Strategies No.1 and 2. The IC method (often referred as the unmodified incomplete factorization method) should therefore, in our opinion, be considered as less robust than the modified incomplete factorization methods. For further comparison between modified and unmodified methods, see Axelsson [2].

6. Conclusion. The major conclusion we can draw from our results is that modified incomplete factorizations of semidefinite systems do not require a special theory or, otherwise stated, that their regular theory does cover the singular case provided that it is written in a formalism which does not exclude singularity by itself. From a practical point of view, it means that a computer program designed for the PCG solution of positive definite systems with preconditioning determined from a modified incomplete factorization can actually also work for semidefinite systems with nearly no changes.

Acknowledgements. I thank Professor R. BEAUWENS for usefull comments and suggestions.

REFERENCES

[1] O. AXELSSON, *Solution of linear equations : iterative methods*, in : Sparse Matrix Techniques (V.A. Barker, Editor), Lectures Notes in Mathematics No. 572, Springer-Verlag, 1977.

[2] ———, *On the eigenvalue distribution of relaxed incomplete factorization methods and the rate of convergence of conjugate gradient methods*, Technical Report, Departement of Mathematics, Catholic University, Nijmegen, The Netherlands, 1989.

[3] O. AXELSSON AND V. BARKER, *Finite Element Solution of Boundary Value Problems. Theory and Computation*, Academic Press, New York, 1984.

[4] O. AXELSSON AND G. LINDSKOG, *On the eigenvalue distribution of a class of preconditioning methods*, Numer. Math., 48 (1986), pp. 479–498.

[5] ———, *On the rate of convergence of the preconditioned conjugate gradient method*, Numer. Math., 48 (1986), pp. 499–523.

[6] R. BEAUWENS, *Upper eigenvalue bounds for pencils of matrices*, Lin. Alg. Appl., 62 (1984), pp. 87–104.

[7] ———, *On Axelsson's perturbations*, Lin. Alg. Appl., 68 (1985), pp. 221–242.

[8] ———, *Lower eigenvalue bounds for pencils of matrices*, Lin. Alg. Appl., 85 (1987), pp. 101–119.

[9] ———, *Approximate factorizations with S/P consistently ordered M-factors*, BIT, 29 (1989), pp. 658–681.

[10] ———, *Modified incomplete factorizations strategies*, submitted for publication in PCG Conference Proceedings, (1990).

[11] R. BEAUWENS AND L. QUENON, *Existence criteria for partial matrix factorizations in iterative methods*, Siam J. Numer. Anal., 13 (1976), pp. 615–643.

[12] R. BEAUWENS AND R. WILMET, *Conditioning analysis of positive definite matrices by approximate factorizations*, J. Comput. Appl. Math., 26 (1989), pp. 257–269.

[13] A. BEN ISRAEL AND T. GREVILLE, *Generalized Inverses : theory and applications*, J. Wiley and Sons, New York, 1974.

[14] A. BERMAN AND R. PLEMMONS, *Nonnegative Matrices in the Mathematical Sciences*, Academic Press, New York, 1979.

[15] A. GEORGE AND J. LIU, *Computer Solution of Large Sparse Positive Definite Systems*, Prentice-Hall, Englewood Cliffs, 1981.

[16] I. GUSTAFSSON, *Modified incomplete Cholesky (MIC) methods*, in : Preconditioning Methods, Theory and Applications (D.J. Evans, Editor), Gordon and Breach Science, 1983.

[17] L. HAGEMAN AND D. YOUNG, *Applied Iterative Methods*, Academic Press, New York, 1981.

[18] F. HARARY, *Graph Theory*, Addison-Wesley, Reading, 1969.

[19] E. KAASSCHIETER, *Preconditioned conjugate gradients for solving singular systems*, J. Comp. Appl. Math., 24 (1988), pp. 265–275.

[20] V. LEBEDEV AND S. FINOGENOV, *Utilisation of ordered Chebyshev parameters in iterative methods*, URSS Comput. Math. Math. Phys., 16 (1976), pp. 70–83.

[21] M. MAGOLU, *Lower eigenvalue bounds for singular pencils of matrices*, in preparation.

[22] Y. NOTAY, *Sur le conditionnement de matrices de Stieltjes par factorisations approchées*, Mémoire. Université Libre de Bruxelles, Brussels, Belgium, 1987.

[23] ———, *Polynomial acceleration of iterative schemes associated with subproper splittings*, J. Comp. Appl. Math., 24 (1988), pp. 153–167.

[24] ———, *Incomplete factorizations of singular linear systems*, BIT, 29 (1989), pp. 682–702.

[25] ———, *Conditioning of Stieltjes matrices by S/P consistently ordered approximate factorizations*, submitted for publication, (1990).

[26] A. VAN DER SLUIS AND H. VAN DER VORST, *The rate of convergence of conjugate gradients*, Numer. Math., 48 (1986), pp. 543–560.

[27] R. VARGA, *Matrix iterative analysis*, Prentice Hall, Englewood Cliffs, 1962.

The convergence behaviour of preconditioned CG and CG-S in the presence of rounding errors

H.A.Van der Vorst
Delft University of Technology
Delft, the Netherlands

September 17, 1990

Abstract

Incomplete Choleski decompositions and modified versions thereof are quite effective preconditioners for the conjugate gradients method. It is explained why Modified Incomplete Choleski may lead to more iteration steps, in some cases, whereas it should theoretically be more effective than standard Incomplete Choleski. It is also shown, by carefully analyzing a numerical example, why the convergence behaviour of Conjugate Gradients-Squared can be, sometimes unacceptable, irregular.

1 Introduction

In [17] it is shown, for a model problem, that Incomplete Choleski preconditioning, when solving $Ax = b$ by the conjugate gradients method (CG), has two major effects. It reduces the condition number of the operator, so that the CG process converges right from the start. The other effect is that the preconditioned operator has larger gaps between extremal eigenvalues at one end of the spectrum. Since larger gaps lead to a faster convergence of CG in the direction of the corresponding eigenvector, this explains the so-called superlinear convergence behaviour of CG [15].

This convergence behaviour of CG is quite well understood if we exclude the influence of rounding errors. In practice, however, rounding errors may have a substantial effect. We will show that they may even spoil a convergence behaviour that would have been superior in exact arithmetic. Our numerical example may help to explain why in some actual computations Modified Incomplete Choleski preconditioning fails to be better than the standard process. This is in obvious disagreement with theory, which predicts that, in exact arithmetic, the modified variant should be much more effective.

The Conjugate Gradients-Squared method [14] may, of course, also suffer from the influence of rounding errors, for the same reasons as CG. But, also by example, we will show that there are other causes for its often quite irregular convergence behaviour. These other causes may result in severe cancellation and loss of accuracy in the iterands. This is quite different from the aforementioned effects of rounding errors, which only have the effect of delaying the convergence of the iteration process.

2 Conjugate gradients

2.1 General properties

Let a linear system $Ax = b$ be given with A a symmetric positive definite matrix. Let x_0 denote an (arbitrary) starting vector, and $r_0 = b - Ax_0$ is the corresponding initial residual vector.

Then the standard conjugate gradient method [8],[6], for solving $Ax = b$, generates a sequence $x_1, x_2, \ldots$ with the following properties:

$$x_i \in x_0 + K_i(A; r_0) \quad , \tag{1}$$

with

$$K_i(A; r_0) = span(r_0, Ar_0, \ldots, A^{i-1}r_0) \tag{2}$$

(the Krylov subspace).

The vector x_i is the argument for which $\| x - u \|_A$ is minimal, over all $u \in x_0 + K_i(A; r_0)$, where $\| v \|_A{}^2 = (v, Av)$.

From (1) it follows for any $u \in x_0 + K_i(A; r_0)$ that

$$x - u = q(A)(x - x_0) \quad , \tag{3}$$

with $q \epsilon \Pi^1{}_i$, where $\Pi^1{}_i$ denotes the class of polynomials of degree at most i and constant term 1.

Then, from the minimization property of x_i, it follows that

$$\| x - x_i \|_A = min \| q(A)(x - x_0) \|_A \quad , \tag{4}$$

where the minimum is taken over all $q \epsilon \Pi^1{}_i$. Corresponding to r_0 there are uniquely determined eigenvalues $\lambda_1 < \lambda_2 < \ldots < \lambda_m$ and normalized eigenvectors $z_1, z_2, \ldots, z_m$ of A such that $r_0 = \sum_1^m \mu_j z_j$, $\mu_j > 0$ for all j. In exact arithmetic only those eigenvalues and eigenvectors are the active ones and then, obviously, $dim K_i(A; r_0) = i$ for $i \leq m$. However, in the presence of rounding errors, also those eigenvalues and eigenvectors which are not present in r_0 may influence the process gradually. Therefore it does not make much sense to study the convergence behaviour of CG for situations in which certain eigenvalues/eigenvectors are excluded.

Let R_i denote the matrix which has as its columns $r_0/ \| r_0 \|, r_1/ \| r_1 \|, \ldots, r_{i-1}/ \| r_{i-1} \|$, then $R_i{}^T A R_i$ can be viewed as a representation of A projected onto $K_i(A; r_0)$. The eigenvalues $\theta_1^{(i)} < \theta_2^{(i)} < \ldots < \theta_i^{(i)}$ of $R_i{}^T A R_i$ will be called the Ritz values of A with respect to $K_i(A; r_0)$.

The Ritz values tend to converge towards eigenvalues of A for increasing i. In practical situations, the matrix $R_i{}^T A R_i$ can be constructed at low costs from the CG-iteration constants, and the Ritz values can be easily computed. This information can be used to study local effects in the convergence behaviour of CG (and also CG-S).

2.2 The convergence behaviour of CG

From (4) it follows that

$$\begin{aligned} \| x_i - x \|_A &\leq \| q(A)(x - x_0) \|_A \\ &\leq \| q(A) \|_2 \| x - x_0 \|_A \\ &\leq max_{x \epsilon [\lambda_1, \lambda_m]} |q(x)| \, \| x - x_0 \|_A \quad , \end{aligned} \tag{5}$$

for any $q \in \Pi^1{}_i$.

If we take for $q(x)$ the Chebychev polynomial shifted to $[\lambda_1, \lambda_m]$ and normalized such that $q(0) = 1$, then we obtain the well-known upperbound [2],[4],[6]:

$$\| x_i - x \|_A \leq 2\left(\frac{\sqrt{c_{m,1}} - 1}{\sqrt{c_{m,1}} + 1}\right)^i \| x_0 - x \|_A \quad , \tag{6}$$

with $c_{m,1} = \lambda_m / \lambda_1$.

In practical situations it turns out that this upperbound is often quite close for moderate values of i. However, since it only predicts linear convergence by a fixed factor $(\sqrt{c_{m,1}} - 1)/(\sqrt{c_{m,1}} + 1)$, it fails to help us understand local effects as well as the so-called superlinear convergence behaviour [4].

All these effects have been studied in [15]. In particular, the increase in the speed of convergence has been explained as follows. As soon as a Ritz value comes close to, say, λ_1, the CG process behaves from then on as a process corresponding to a starting vector in which z_1 (and λ_1) is not present. The latter process converges with a reduction factor $(\sqrt{c_{m,2}}-1)/(\sqrt{c_{m,2}}+1)$, i.e., we have, in average, faster convergence from then on.

In [15] the observation was made that only a modest degree of approximation of $\theta_1^{(i)}$ to λ_1 suffices for the process to shift to a lower reduction factor. E.g., it suffices already when $\theta_1^{(i)} < (\lambda_1 + \lambda_2)/2$.

This approach implies that we have to study the Ritz values in order to understand the CG-convergence behaviour. It is well-known that $\theta_1^{(i)}$ converges monotonically, for increasing i, towards λ_1 and it has been observed [12] that the speed of convergence is larger for a larger gap ratio $\frac{\lambda_2 - \lambda_1}{\lambda_m - \lambda_1}$.

Also more local effects in the convergence behaviour of CG could be related to the location of the Ritz values for successive values of i. Many of such irregularities have been explained in [15].

2.3 Preconditioning of CG

For a suitably chosen nonsingular matrix $K = LL^T$ we can also apply CG to solve the system

$$L^{-1}AL^{-T}y = L^{-1}b \quad , \quad x = L^{-T}y \quad . \tag{7}$$

This approach is called preconditioning. For a practical conjugate gradient algorithm, involving preconditioning, see, e.g., [6], [10], [18].

In view of the preceeding section, effective preconditioning should lead to a matrix $L^{-1}AL^{-T}$ which has either a smaller condition number or larger gap ratios for extremal eigenvalues (or both), with respect to A. By an effective preconditioner we mean that it should lead to a considerable reduction in the number of iteration steps.

In [17] it is shown for a model problem, the discretized Poisson equation, that *Incomplete Choleski* preconditioning [10] has both effects. Though the reduction of the condition number is significant, it is not by an order of magnitude and it can be shown that $L^{-1}AL^{-T}$, as well as A, has extremal eigenvalues $\lambda_1 = O(h^2)$, $\lambda_m = O(1)$, so that the condition number is $O(h^{-2})$, where h denotes the discretisation step. The preconditioned CG method is often referred to as ICCG.

Modified Incomplete Choleski preconditioning [7] has a different effect, with respect to Incomplete Choleski, insofar that $\lambda_1 = O(1)$, $\lambda_m = O(h^{-1})$ and hence the condition number is reduced to $O(h^{-1})$. Moreover, the largest eigenvalues have quite large gap ratios. At least in exact arithmetic, one would expect Modified ICCG, or MICCG, to be much more effective than the standard ICCG process, and, indeed, in actual computation this turns out to be the case for many problems. However, for many other realistic situations it happens that MICCG leads to an even slower iteration process than ICCG. In section 3 we will give an example of this, along with an explanation for the observed effects.

3 A numerical example for preconditioned CG

All the experiments reported in this section have been carried out in 48-bit floating point arithmetic on CRAY computers. Some of these computations have also been done on other computers as well, leading to the same phenomena in a qualitative sense. In other words, if one carries out similar experiments on any given computer, then one will observe similar, but of course not exactly the same, effects as those that we report here.

The in section 2 presented theory on the convergence behaviour of CG does not take into account the effects of rounding errors. These may have a profound effect as we will see in this section.

In our example the linear system arises from standard 5-point finite difference discretization, over an equidistant rectangular grid, of the following pde:

$$-\frac{\partial}{\partial x}(A\frac{\partial u}{\partial x})-\frac{\partial}{\partial y}(A\frac{\partial u}{\partial y})=F \quad , \tag{8}$$

over the unit square.
The boundary conditions are:
$u=0$, for $0\leq x\leq 1$, $y=0$,
$\frac{\partial u}{\partial n}=0$ along the rest of the boundary.
The functions A and F are given by:
$A=100$, $F=100$ for $0.25\leq x\leq 0.75$ and $0.25\leq y\leq 0.75$,
$A=1$, $F=0$ over the remaining part of the region.
The grid has a mesh spacing of $\frac{1}{32}$, so that a linear system of order $33*32=1056$ results.

In Table I we have given the numbers of iterations, for both *ICCG* and *Modified ICCG*, required in order to get a residual $||Ax_i-b||_2$ below a specified value of ε.

Table I. Iteration results for ICCG and MICCG

ε	number of iterations for ICCG	MICCG
10^{-1}	21	18
10^{-2}	26	26
10^{-3}	31	31
10^{-4}	35	35
10^{-5}	40	42
10^{-7}	49	51
10^{-9}	54	61

In view of the results in [15] we make the following comments:

1. For this example the condition number associated with MICCG is indeed far less than the one associated with ICCG: 214.8 and 5444, respectively. Therefore, we may expect MICCG to be at least faster in the early iteration phase.

2. After a number of iterations the effective condition numbers for both processes are more or less the same, e.g., $\simeq 12$ for MICCG and $\simeq 19$ for ICCG, after 40 iteration steps.

3. In contrast to what we might expect then, ICCG is even more efficient than MICCG after 40 iteration steps.

By inspecting the associated Ritz values for each of the processes, we soon discover why the convergence behaviour of MICCG is spoiled in this case.

The largest eigenvalues associated with MICCG are very well separated so that they are approximated almost up to machine precision by Ritz values in an early stage of the iteration process. As soon as a Ritz value has converged up to machine precision then the following residual vectors can no longer be made orthogonal with respect to the eigenvectors corresponding to the 'converged' eigenvalues [11],[12]. It is very likely (and it happens indeed) that in the new residual vectors components in those 'converged eigenvector directions' are built up because of rounding errors. This means that the orthogonality between these new residual vectors and the older ones is gradually lost and earlier discovered Ritz values are discovered again: the Ritz values appear as multiplets (this is in good agreement with [12]:chapter 13). Note that this loss of orthogonality will take place on any computer and in any finite arithmetic system, be it that if we use a shorter or longer mantissa the loss will occur earlier or later. This loss can not be prevented by, e.g., computing the residual vectors more accurately. As we said before, we have run these experiments on different computers and in different precisions and they led all to qualitatively the same results

As an illustration, we give the five largest different eigenvalues associated with MICCG and we list between parenthesis the multiplicity of the corresponding Ritz values after 61 steps:

$\lambda_1 = 214.82705 \quad (7*)$ $\qquad \lambda_4 = 35.062177 \quad (3*)$
$\lambda_2 = 57.436865 \quad (4*)$ $\qquad \lambda_5 = 27.363301 \quad (2*)$
$\lambda_3 = 48.555435 \quad (4*)$

We note that none of the smallest eigenvalues associated with ICCG is represented by a multiple Ritz value, after 54 iteration steps. For ICCG it are the smallest eigenvalues that determine the convergence behaviour, see [17].

The presence of the multiplets for MICCG hampers its convergence behaviour. Namely, in a phase of the iteration process that a new member of a multiplet gets born, the effective condition number increases accordingly, as can be seen by inspecting the convergence behaviour in that phase. If we make, after convergence of a Ritz value, the successive residual vectors explicitly orthogonal with respect to the 'converged' eigendirection, then the number of iteration steps drops for MICCG and its convergence behaviour becomes in good agreement with the theory developed in [15].

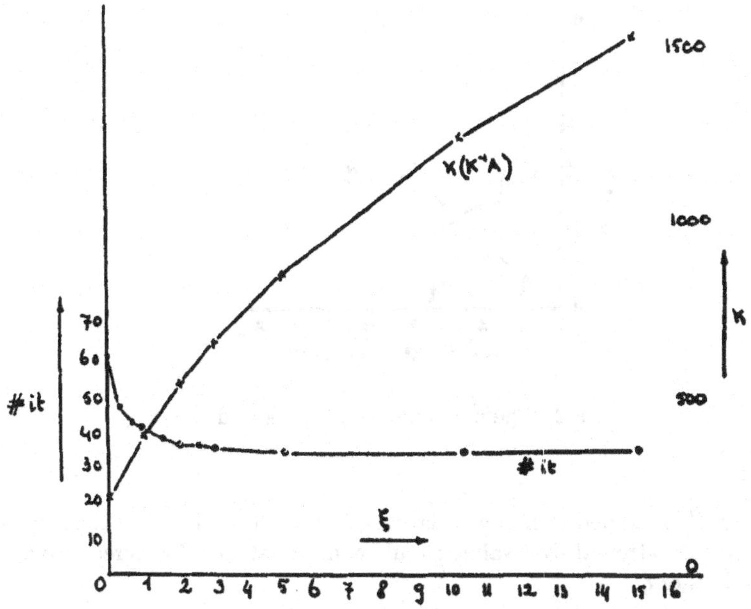

Figure 1. The effect of ξ on MICCG

We have done this explicit reorthogonalization in high precision only in order to verify that it is indeed the loss of orthogonality that detoriates the convergence behaviour of MICCG. We do not suggest that our procedure would be very practical. In fact, we have not considered the problem whether there exists an efficient way of reorthogonalizing, since we believe that it is much simpler to follow the strategy that we will describe now.

In Modified Incomplete Choleski corrections are added to the main diagonal of L (and L^T), such that the rowsums of LL^T are equal to the corresponding rowsums of A. It has been suggested by many authors, e.g., see [1],[3],[18], to make only slightly smaller corrections, 95% say, and they report a significant decrease in the number of iterations for MICCG in many cases. The smaller corrections have the same effect as making a Modified Incomplete Choleswki decomposition of $A + \xi h^2$ and taking that as the preconditioner K. In Figure 1 we have plotted the condition number of $K^{-1}A$, as well as the number of iterations required to reduce the initial residual by a factor of 10^{-9}, versus the value of ξ. Again we observe the remarkable and, at first sight, contradictory effect that $\kappa(K^{-1}A)$ increases, while, for small values of ξ, the number of iteration steps (sharply) decreases.

As before, these effects can be explained by losses of orthogonality. What we have seen in our experiments is that the extremal eigenvalues are very well separated for $\xi = 0$, so that we have an early (and repetitive) loss of orthogonality. For increasing values of ξ, the extremal eigenvalues are gradually less well separated and while $\kappa(K^{-1}A)$ increases, the reduced loss of orthogonality more than compensates for the theoretically expected increase in the number of iteration steps.

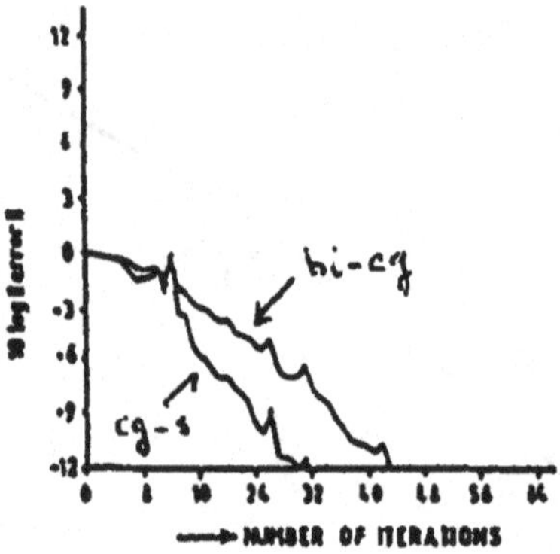

Figure 2. The convergence of bi-cg and CG-S

We have seen this happen in many situations, but we have no proof that the smaller gap ratios for the extremal eigenvalues should compensate for the increase in $\kappa(K^{-1}A)$ for small values of ξ.

4 Conjugate gradients-squared (CG-S)

4.1 Background of the method

The residual vectors r_i, generated by the CG method, satisfy a three- term recurrence relation. This is fundamental for the efficiency of CG. When A is not symmetric we loose this property for the r_i's. Also, when A is not positive definite then $\| \ \|_A$ does not define a norm, so that it does not make sense to minimize $\| x - x_i \|_A$.

In the *Bi-Conjugate Gradient* method [5], the approximations are constructed in such a way that the residual r_j is orthogonal with respect to another row of vectors $\hat{r}_0, \hat{r}_1, \ldots, \hat{r}_{j-1}$, and, vice versa, $\hat{r}_j$ is orthogonal with respect to $r_0, r_1, \ldots, r_{j-1}$. This can be accomplished by two 3-term recurrence relations for the rows $\{r_j\}$ and $\{\hat{r}_j\}$. The Bi-Conjugate Gradient method also terminates within n steps at most, but there is no minimization property as in CG for the intermediate steps.

Sonneveld [14] made the observation that, in case of convergence, both the rows $\{r_j\}$ and $\{\hat{r}_j\}$ converge towards zero, but that only the convergence of the $\{r_j\}$ is exploited. He proposes a modification to Bi-Conjugate Gradients by which all the convergence effort is concentrated in the r_i vectors. It turns out that this modified method just generates residual vectors r_i that can be written as

$$r_i = P^2{}_i(A) r_0 \quad , \tag{9}$$

in which $P_i(A)$ is the ith degree polynomial in A, such that $P_i(A)r_0$ is the Bi-Conjugate Gradients residual at the ith iteration step.

Hence, if $P_i(A)$ is viewed as a reductor working on r_0, then this reductor is now applied twice in the new method and this explains the name *Conjugate Gradients-*

Squared (CG-S). The prefix "bi-" has been omitted since there are no longer two bi-orthogonal rows $\{r_j\}$ and $\{\hat{r}_j\}$.

Note that when A is symmetric positive definite then the CG-S polynomial $P^2{}_i(A)$ is just the square of the ith degree minimizing polynomial in CG. We will use this property in our discussion on the convergence behaviour of CG-S for a simple model problem.

4.2 The convergence behaviour of CG-S

A typical iteration history is given in Figure 2, which has been taken from [9]. It has been observed in many situations where Bi-Conjugate Gradients converges that CG-S converges about twice at fast (at about the same computational costs per iteration step). Similarly, when Bi-Conjugate Gradients diverges, then CG-S diverges about twice as fast. This behaviour could be expected, heuristically, by taking the viewpoint that the Bi-Conjugate Gradients operator $P_i(A)$ is applied twice to r_0 in CG-S.

The weak point in this way of reasoning is that the reduction operator $P_i(A)$ in Bi-Conjugate Gradients is very dependent on the starting initial residual r_0, and that it is not likely to be a reduction operator for any other vector, even not for the particular vector $P_i(A)r_0$ itself. Indeed, it is not very difficult to construct examples for which $P_i(A)r_0$ is small in norm and for wich $P^2{}_i(A)r_0$ is even much bigger in norm.

Such a phenomenon may happen in the following situation. Suppose that r_0 has small coordinates in some eigenvector directions of the matrix A. Then $P_i(\lambda)$ may take very large values in the corresponding eigenvalues, in particular when these eigenvalues are more or less isolated from he others, without leading to a significant contribution of that eigenvector direction to the norm of r_i in Bi-Conjugate Gradients. However, the value of $P^2{}_i(\lambda)$ may be so large that the contribution of the corresponding eigenvector direction even dominates in the residual r_i of CG-S.
We will follow this situation in more detail for a simple numerical example.

5 A numerical example for CG-S

The numerical experiments in this section have been obtained by double precision computation on an IBM 3090, but we have obtained similar effects on other machines (and in other precisions).

For a relatively simple problem we will study the convergence behaviour of preconditioned CG-S. From the convergence behaviour of Ritz values we know that a specific eigenvalue is 'discovered' in the Krylov subspace, as soon as the Ritz polynomial [16] is becoming to large in that eigenvalue. Hence we might expect that at this 'point of discovery' [12] the value of the square of this polynomial may be very large with respect to its value in other eigenvalues.

To illustrate this we apply CG and CG-S to a symmetric positive definite system and we use Modified Incomplete Choleski [7] as the preconditioner in both cases. Since Modified Incomplete Choleski may lead to rather big gap ratios for the extremal eigenvalues, we may expect the convergence behaviour of CG-S to be more irregular as for Incomplete Choleski (this was observed in many practical situations to be the

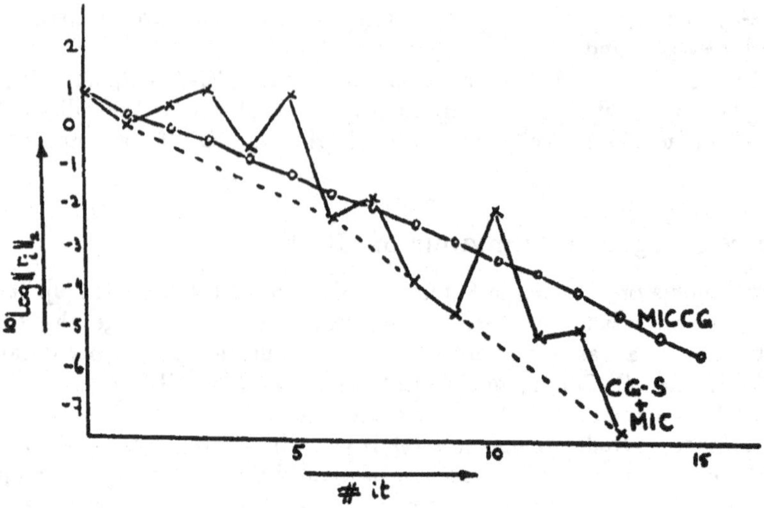

Figure 3. The convergence of preconditioned CG and CG-S

case).

The linear system comes from the discretized Poisson equation

$$\Delta u = f$$

over the unit square, with Dirichlet boundary conditions. A uniform $30 \star 30$ grid has been imposed so that a matrix of order $28 \star 28$ results.

In Figure 3 we have plotted the convergence history for preconditioned CG and CG-S. If we connect the minima in the curve for CG-S then, with some phantasy, we note a quadratic onvergence behaviour with respect to the curve for CG. But we see also that there are many local peaks in the convergence curve for CG-S. Note that these peaks do not seem to delay the convergence of CG-S. However, in practical situations they may have quite an adverse effect, since they may be so large that the local corrections to the current iterate result in cancellation. As a result the final solution may have little significance, which can be checked by computing the real residual (instead of the updated residual as is done usually in these processes).

In Figure 4 we have also plotted the Ritz values for CG for successive iteration steps. Each sequence of Ritz values, for increasing iteration count i, that seems to converge to a specific eigenvalue, has been connected, but mathematically speaking, there is not much relation between these Ritz values.

As has been argued before, each time when the Ritz polynomial takes a large value in one specific eigenvalue, the Krylov subspace iteration process starts to locate Ritz values near that eigenvalue. This situation is called the *point of discovery* in [13]. In Figure 4 these situations have been indicated by drawing a new branch connecting

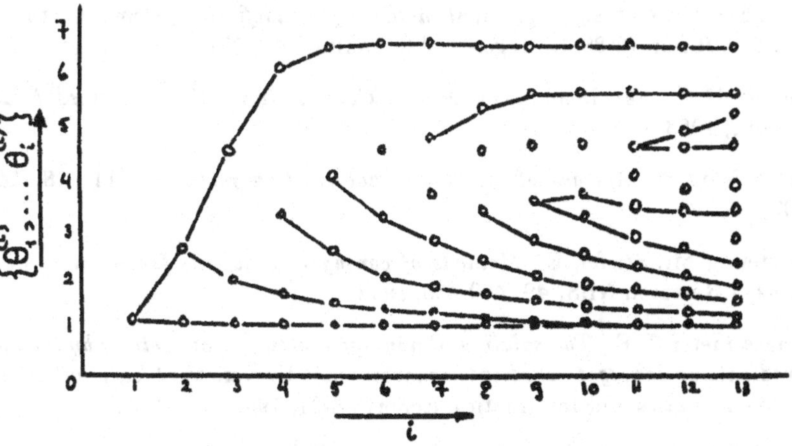

Figure 4. The location of the Ritzvalues in successive iteration steps of CG

the converging subrow of Ritz values.
By comparing Figure 3 and 4, we make the interesting observation that CG-S has a peak in its convergence behaviour at each new point of discovery. This is in nice agreement with our heuristical arguments.

Acknowledgements

I wish to thank the Royal Dutch Meteorological Office (KNMI) for providing me access to the CRAY computer system of ECMWF at Reading, and I am very grateful to Philips Eindhoven for their permission and help in using their IBM 3090 system at Natlab.

References

[1] Ashcraft, C., Grimes, R., *On vectorizing incomplete factorizations and SSOR preconditioners*, SIAM J.Sci.Statist.Comput., 9, 122-151, 1988

[2] Axelsson, O., *Solution of linear systems of equations: iterative metods*, in: Barker, V.A. (ed.), *Sparse Matrix Techniques*, Copenhagen, August 9-12, 1976, Lecture Notes in Mathematics 572, Springer Verlag, Berlin, 1977

[3] Axelsson, O., Lindskog, G., *On the eigenvalue distribution of a class of preconditioning methods*, Numer.Math., 48, 479-498, 1986

[4] Concus, P., Golub, G.H., O'Leary, D.P., *A generalized conjugate gradient method for the numerical solution of elliptic partial differential equations*, in: *Sparse Matrix Computations* (J.R.Bunch and D.J.Rose, eds), pp.309-332, Academic Press, New York, 1976

[5] Fletcher, R., *Conjugate gradient methods for indefinite systems*, Lecture Notes Math. **506**, pp.73-89, Springer-Verlag, Berlin, etc., 1976

[6] Golub, G.H., Van Loan, C.F., *Matrix Computations*, North Oxford Academic, Oxford, 1983

[7] Gustafsson, I., *A class of 1:st order factorization methods*, BIT, **18**, 142-156, 1978

[8] Hestenes, M., Stiefel, E., *Methods of conjugate gradients for solving linear systems*, J.Research NBS, **49**, 409-436, 1952

[9] Kaasschieter, E.F., *The solution of non-symmetric linear systems by bi-conjugate gradients or conjugate gradients squared*, Delft University of Technology, Faculty of Mathematics and Informatics, Report 86-21, 1986

[10] Meijerink, J.A., Van der Vorst, H.A., *An iterative solution method for linear systems of which the coefficient matrix is a symmetric M-matrix*, Math.of Comp., **31**, 148-162, 1977

[11] Paige, C.C., *The computation of Eigenvalues and Eigenvectors of Very Large Sparse Matrices*, Ph.D.thesis, Univ. of London, 1971

[12] Parlett, B.N., *The symmetric eigenvalue problem*, Prentice Hall, Englewood Cliffs, 1980

[13] Parlett, B.N., *Misconvergence in the Lanczos algorithm*, Department of Mathematics, Univ.of California at Berkeley, Report PAM-404, 1987

[14] Sonneveld, P., *CGS: a fast Lanczos-type solver for nonsymmetric linear systems*, SIAM J.Sci.Statist.Comput., **10**, 36-52, 1989

[15] Van der Sluis, A., Van der Vorst, H.A., *The rate of convergence of conjugate gradients*, Numer.Math., **48**, 543-560, 1986

[16] Van der Sluis, A., Van der Vorst, H.A., *The convergence behavior of Ritz values in the presence of close eigenvalues*, Lin.Alg.& its Appl, **88/89**, 651-694, 1987

[17] Van der Vorst, H.A., *Preconditioning by incomplete decompositions*, Ph.D.Thesis, Univ. of Utrecht, 1982

[18] Van der Vorst, H.A., *High performance preconditioning*, SIAM J.Sci. Statist. Comput., **10**, 1174-1185, 1989

DATA REDUCTION (DARE) PRECONDITIONING FOR GENERALIZED CONJUGATE GRADIENT METHODS

R. WEISS, W. SCHÖNAUER
Rechenzentrum der Universität Karlsruhe
Postfach 6980
D-7500 Karlsruhe 1
Federal Republic of Germany

Abstract

The preconditioning of generalized conjugate gradient methods by means of data reduction (DARE) methods for linear systems arising from the finite difference method will be presented. DARE is a multilevel method where unknowns are dropped by using a functional approach. On each level iterations of generalized CG methods are performed according to different strategies. The better the functional approach fits the solution, i.e. the finer the size of the grid compared to the solution curvature, the better DARE works. Numerical experiments show that one gets a nearly linear increase in computation with the number of unknowns and that the convergence of DARE is rather independent from the operator of the partial differential equation. DARE turns out to be a promising and feasible technique, but a lot of work still has to be done looking for optimal strategies.

1 Introduction

We are concerned with the solution of 2-D and 3-D elliptic and parabolic systems of partial differential equations. They are solved numerically by the finite difference method of arbitrary consistency order by the black-box solver FIDISOL, see section 17 in [3]. The matrix resulting from the discretization and linearization of the system on a regulary structured grid is large, sparse and it has the structure of scattered diagonals. Therefore the linear system is solved by generalized CG methods because they are optimally suited for the solution of this type of systems on vector computers [4]. As the solution of the linear system is the main expense of the whole problem solution and consumes about 90 % of the computer time a good preconditioner is searched to accelerate the CG method. A preconditioner based on a multilevel method but different to multigrid methods [1] will be presented in the following. Our aim is to solve the linear system

$$Ax = b \tag{1}$$

where A is usually non-symmetric and not positive definite.

The matrix A results from a black-box solver that handles arbitrary differential operators with arbitrary boundary conditions. The consistency order of the finite difference method is selfadapted or chosen by the user. Therefore, nothing is known of the inner and the outer structure of the matrix. The task has been to develop a robust multilevel method that is applicable under such difficult conditions in contrast to a special multilevel method designed for a special differential equation with special boundary conditions and with a fixed consistency order.

The generalized CG methods have been chosen as solver because of their excellent vectorization properties in connection with the appropriate data structures. There is no closed theory for the different members of the CG family for non-symmetric, non-positive definite matrices. Consequently there is up to now no theory for DARE. But the promising numerical results of DARE shown in the following should encourage to more theoretical research.

2 The Generalized CG methods

Our investigations have been based on generalized CG methods.

Let be x_0 any initial guess for the solution of the system Ax=b. The following recurrence is called pseudo-residual method (PRES).
Choose an "appropriate" d_k

$$r_0 = Ax_0 - b\,, \tag{2}$$

$$\bar{r}_{k+1} = \bar{A}d_k + \sum_{i=1}^{q} \alpha_{i,k} r_{k+1-i}\,, \tag{3}$$

where the $\alpha_{i,k}$ are determined from $\bar{r}_{k+1}^T \bar{r}_{k+1} = min!$, i.e.

$$\alpha_{i,k} = -\frac{r_{k+1-i}^T \bar{A} d_k}{r_{k+1-i}^T r_{k+1-i}}\,, \tag{4}$$

$$r_{k+1} = \phi_k \bar{r}_{k+1}\,, \tag{5}$$

$$x_{k+1} = \phi_k (r_k + \sum_{i=1}^{q} \alpha_{i,k} x_{k+1-i}) \tag{6}$$

with

$$\phi_k = \frac{1}{\sum_{i=1}^{q} \alpha_{i,k}}\,. \tag{7}$$

q is called order of the PRES method.
The quantities $\bar{r}_k$ are called pseudo-residuals. They have the same direction as the residuals but a different length. The minimum condition for the pseudo-residuals induces the orthogonalities $r_{k+1}^T r_{k+1-i} = 0$ for i=1,...,q. The definition for the PRES

methods includes the classical CG method, the Biconjugate Gradient method, the ATPRES method [5] and all variants of the Orthores method [2].

As the norm of the residuals may oscillate heavily we introduce parallel to the original sequence the smoothed sequence

(8) $$s_0 = r_0 \,, x_0^s = x_0 \,,$$

(9) $$s_{k+1} = s_k + \gamma_k(r_{k+1} - s_k) \,,$$

(10) $$x_{k+1}^s = x_k^s + \gamma_k(x_{k+1} - x_k^s)$$

where γ_k is determined from $s_{k+1}^2 = min!$, i.e.

(11) $$\gamma_k = -\frac{s_k^T(r_{k+1} - s_k)}{(r_{k+1} - s_k)^2} \,.$$

The smoothed sequence is a norm non-increasing function of the iteration index k.

We apply two methods for DARE preconditioning.

1. PRES20, where $\tilde{A} = A$, $d_k = r_k$ and q=min(k+1,5). After 20 iterations the sequence is restarted from the smoothed sequence. This method is fast for well-conditioned matrices but not very robust.

2. ATPRES, where $\tilde{A} = A$, $d_k = A^T r_k$ and q=2 as the sequence terminates automatically. This method is slow but very robust. This method can be considered as preconditioning by A^T from right.

3 Preconditioning

3.1 Preconditioning from the left-hand side

For all iterative methods it is essential to normalize the linear system , i.e. to balance the equations. We divide each row of the matrix and the right-hand side by the positive or negative sum of the absolute values of the elements in this row. The sign is choosen so that all main diagonal entries have a positive sign. In the following this preconditioning from the left-hand side is always assumed.

3.2 Preconditioning from the right-hand side

In the terms of the PRES algorithm preconditioning from the right-hand side means to choose a good d_k. The optimal d_k would be the solution of

(12) $$Ad_k = r_k$$

as the exact solution of system (1) would be obtained in the first iteration step. But the solution of the system (12) is the same problem as the solution of the original system (1). Therefore the system (12) is solved approximately by the DARE method. The resulting d_k is used as search direction in the "leading" PRES method.

Figure 1: Linar case

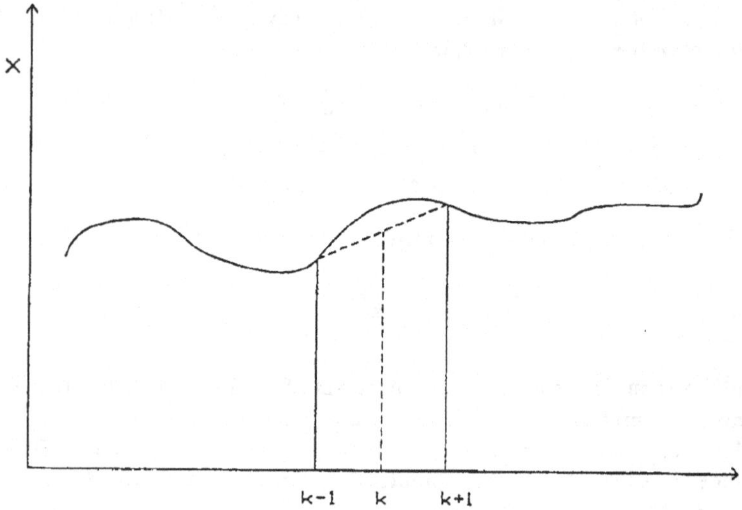

4 The basic idea of DARE

The basic idea of DARE is to drop unknowns and equations and to modify the resulting smaller system according to a functional approach for the dropped unknowns. We have been investigating a linear, quadratic and cubic dependence of the unknown x_k on the index k. In the linear case the unknown x_k is assumed to be $x_k = 0.5(x_{k-1}+x_{k+1})$, see Fig. 1. In the equation i the coefficient $a_{i,k}$ is distributed onto the neighbouring coefficients $a_{i,k-1}$ and $a_{i,k+1}$ with the weights 0.5.

In the cubic case the unknown x_k is assumed to lie on the cubic polynomial which fits the unknowns $x_{k-3}, x_{k-1}, x_{k+1}$ and x_{k+3}, i.e.

$$x_k = \alpha_1 x_{k-3} + \alpha_2 x_{k-1} + \alpha_3 x_{k+1} + \alpha_4 x_{k+3} \tag{13}$$

with $\alpha_1 + \alpha_2 + \alpha_3 + \alpha_4 = 1$ (see Fig. 2).

In the equation i the coefficient $a_{i,k}$ is distributed onto the neighbouring coefficients $a_{i,k-3}, a_{i,k-1}, a_{i,k+1}$ and $a_{i,k+3}$ according to the weights α_l for $l = 1, \ldots, 4$.

Now we drop every second unknown x_k and every second equation. The result is a linear system of half size.

We continue this procedure until we end up at a 2x2 system or a 3x3 for linear and a 4x4 or 5x5 system for cubic approach (the treatment of end variables is explained below). This system will be solved by the direct Gauss method. The solution of this system will be extended to an initial guess for the solution of the next finer system by interpolating the dropped unknowns according to the functional approach.

Then we perform some iterations of the PRES method on the finer grid until a desired accuracy is obtained or the method "falls asleep". This procedure is continued till we end up at the finest (full) system. The approximate solution d_k of this system is taken as search direction for the governing PRES method (see Fig. 3 for linear approach).

Figure 2: Cubic case

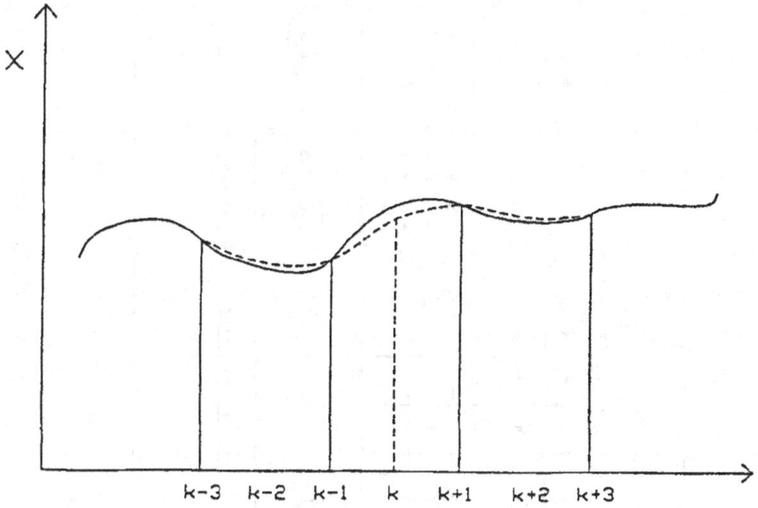

The algebraic formulation of DARE is the following. Let nlevel be the number of reduction steps and let be

$$A_0 = A, \quad b_0 = r_k \; . \tag{14}$$

Consider the sequence of linear systems for $i = 0, \dots, nlevel$

$$A_i x_i = b_i \tag{15}$$

$$with$$

$$A_i = D_{i-1} A_{i-1} W_{i-1}, \quad b_i = D_{i-1} b_{i-1} \; . \tag{16}$$

D_i is the matrix that eliminates the rows of the matrix A_i, and W_i is the weight matrix that distributes the entries of the matrix A_i so that the columns are deleted. Let $\bar{x}_{i+1}$ be the approximate solution of the system of level i+1. Choose as starting guess for the system on the level i the vector

$$x_i = W_i \bar{x}_{i+1} \; . \tag{17}$$

As search direction for the original system we use $d_k = \bar{x}_0$.

For the 1-D reduction with $n = 2^q + 1$ grid points the matrix $D_i \in I\!R^{(2^{q-i-1}+1) \times (2^{q-i}+1)}$ deletes each second row:

$$D_i = \begin{pmatrix} 1 & 0 & \dots & \dots & \dots & \dots & \dots & \dots & \dots & \dots & \dots & 0 \\ 0 & 0 & 1 & 0 & \dots & \dots & \dots & \dots & \dots & \dots & \dots & 0 \\ 0 & 0 & 0 & 0 & 1 & 0 & \dots & \dots & \dots & \dots & \dots & 0 \\ \vdots & & & & & & & & & & & \vdots \\ 0 & \dots & \dots & \dots & \dots & \dots & \dots & \dots & \dots & 1 & 0 & 0 \\ 0 & \dots & \dots & \dots & \dots & \dots & \dots & \dots & \dots & \dots & 0 & 1 \end{pmatrix} . \tag{18}$$

Figure 3: DARE - reduction and interpolation, linear approach

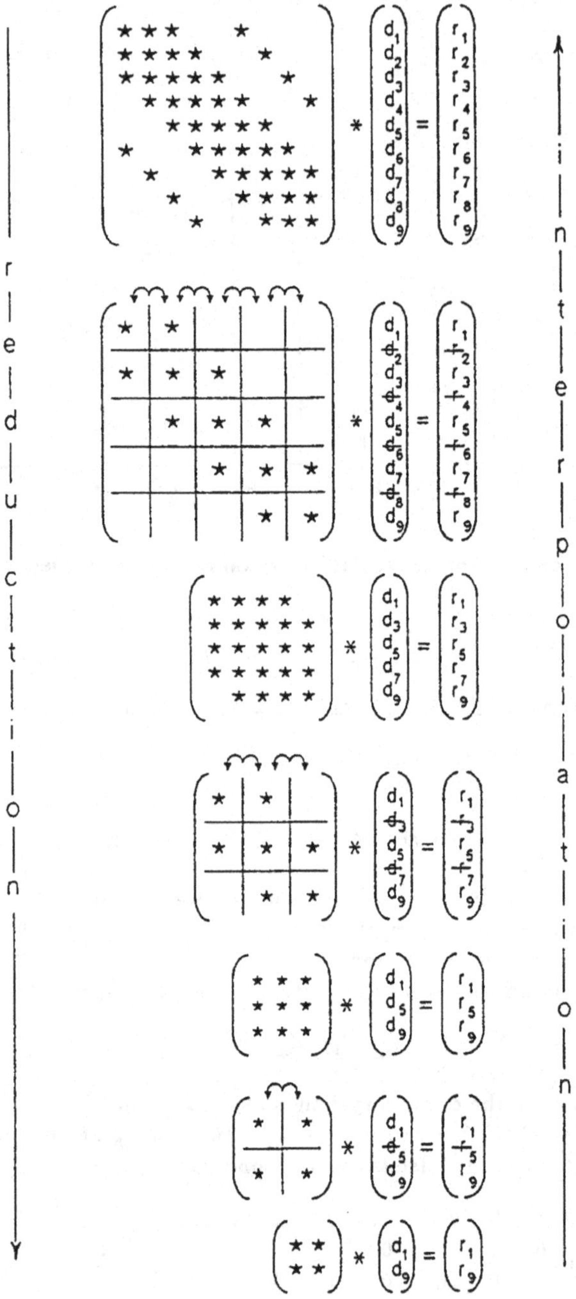

The weight matrix $W_i \in I\!R^{(2^{q-i}+1) \times (2^{q-i-1}+1)}$ consists of the weights α_i. For the linear case we get

$$
(19) \qquad W_i = \begin{pmatrix}
1 & 0 & \dots & \dots & \dots & 0 \\
0.5 & 0.5 & 0 & & & 0 \\
0 & 1 & 0 & 0 & \dots & 0 \\
 & 0.5 & 0.5 & 0 & & 0 \\
0 & 0 & 1 & 0 & & 0 \\
0 & 0 & 0.5 & 0.5 & & 0 \\
\vdots & & & & & \vdots \\
0 & & & & 0.5 & 0 \\
0 & & & & 1 & 0 \\
0 & & & & 0.5 & 0.5 \\
0 & \dots & \dots & \dots & 0 & 1
\end{pmatrix}.
$$

For the cubic case we get at the end points the different weights $\alpha_{1a}, \alpha_{2a}, \alpha_{3a}, \alpha_{4a}, \alpha_{1b}, \dots, \alpha_{4b}, \alpha_{1c}, \dots, \alpha_{4c}, \alpha_{1d}, \dots, \alpha_{4d}$ resulting in the following weight matrix:

$$
(20) \qquad W_i = \begin{pmatrix}
1 & 0 & \dots & \dots & \dots & \dots & \dots & \dots & 0 \\
\alpha_{1a} & \alpha_{2a} & \alpha_{3a} & \alpha_{4a} & 0 & & & & 0 \\
0 & 1 & 0 & \dots & \dots & \dots & \dots & \dots & 0 \\
\alpha_{1b} & \alpha_{2b} & \alpha_{3b} & \alpha_{4b} & 0 & & & & 0 \\
0 & 0 & 1 & 0 & 0 & & & & 0 \\
\alpha_1 & \alpha_2 & \alpha_3 & \alpha_4 & 0 & & & & 0 \\
0 & 0 & 0 & 1 & 0 & & & & 0 \\
0 & \alpha_1 & \alpha_2 & \alpha_3 & \alpha_4 & & & & 0 \\
0 & 0 & 0 & 0 & 1 & & & & 0 \\
\vdots & & & & & & & & \vdots \\
0 & & & & & \alpha_1 & \alpha_2 & \alpha_3 & \alpha_4 \\
0 & & & & & 0 & 1 & 0 & 0 \\
0 & & & & & \alpha_{1c} & \alpha_{2c} & \alpha_{3c} & \alpha_{4c} \\
0 & & & & & 0 & 0 & 1 & 0 \\
0 & & & & & \alpha_{1d} & \alpha_{2d} & \alpha_{3d} & \alpha_{4d} \\
0 & \dots & \dots & \dots & \dots & \dots & \dots & 0 & 1
\end{pmatrix}.
$$

In the following the DARE technique will be generalized for an arbitrary number of grid points, for 2-D and 3-D problems and for systems of partial differential equations. For these problems the matrices D_i and W_i are more sophisticated. As we never use these matrices explicitly for the DARE technique and as the structure of these matrices is pretty complex we do not indicate their structure in the following. Note that $D_i^T \neq W_i$ in contrast to most multigrid methods.

As the cubic DARE procedure turned out to be superior to the linear and quadratic case [6,7] we only use the cubic version. But for simplification we explain the main features of DARE using the linear aproach in this paper.

One preconditioned PRES step is called one cycle. The time proceeding is depicted in Fig. 4.

Figure 4: DARE procedure

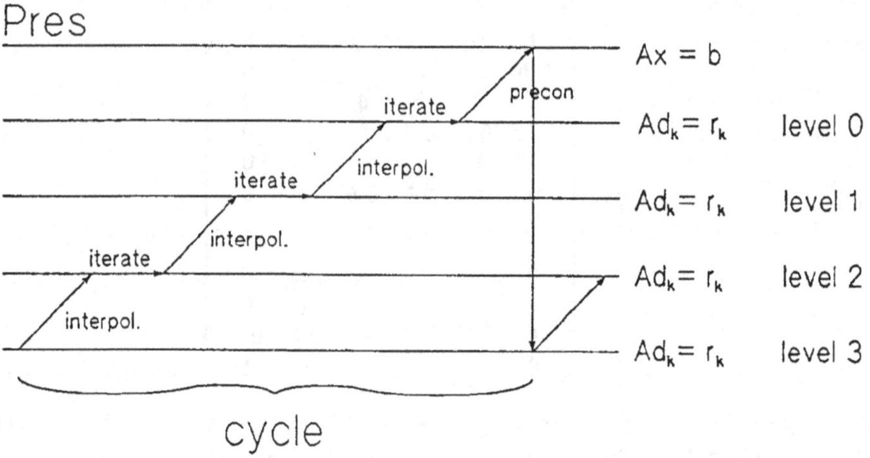

5 Information transfer

By the DARE method the global information is introduced into the CG solution procedure much cheaper than by the CG method itself. The information of the boundary points is transfered to the interior of the domain by the interpolation as shown in Fig. 5.

The better the functional approach fits the solution the less expensive iterations are required on fine grids. The methods works the better, the finer the grid size is, compared to the solution curvature, i.e. the smoother the solution is compared to the grid size.

Figure 5: Information transfer

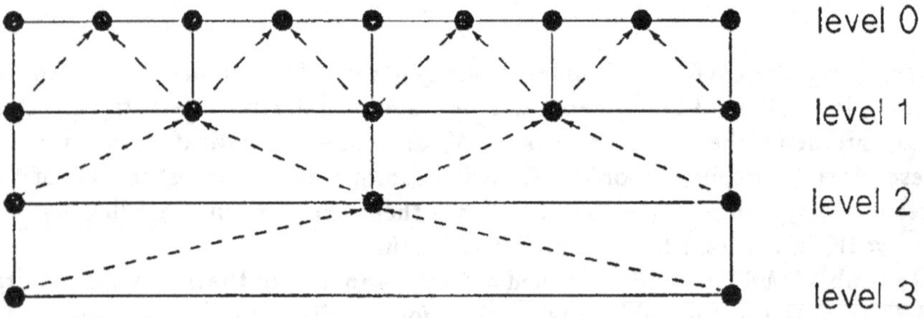

Figure 6: Reduction for n=16

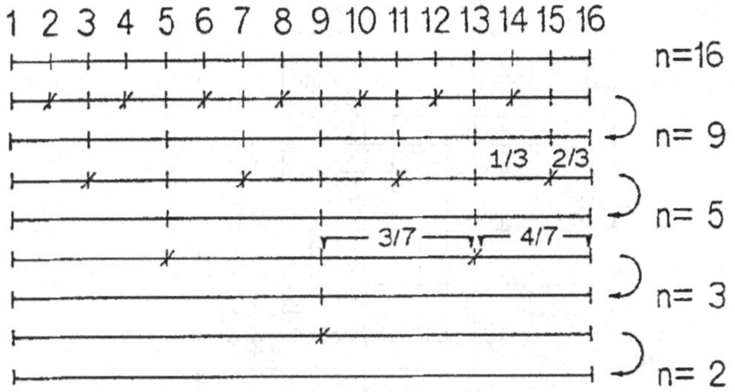

6 The number of grid points

Up to this point we assumed the number of grid points n to be $n = 2^q + 1$ for an arbitrary integer q. But as we didn't want to be confined to this special choice we had to introduce a special elimination process for an arbitrary n. The end points of a grid are always kept and for an even n the two last points on the right end and thus the two last variables are kept which requires different weights in the next reduction steps as indicated in Fig. 6 for n=16. This is the procedure in 1-D problems.

7 2-D and 3-D reduction

As the problems arise from differential equations on 2-D and 3-D domains an alternating reduction and interpolating scheme for different directions is applied. In the case of a y- or z-reduction and for lexicographical order of the grid points this requires a block elimination for lines with constant x or planes with constant x and y (see Fig. 7 for the y-elimination on a 3x5 grid).

8 Reduction for systems of equations

Up to now the DARE method was discussed for a single partial differential equation. In the usual case we have a problem with N partial differential equations and N different variables representing different physical properties as for example velocities u, v and pressure p. Of course it is useless to reduce velocities depending on pressure or to interpolate pressure from the velocity. The DARE method has to distinguish different physical variables. As the discretized variables are arranged in the form $u_1, v_1, p_1, u_2, v_2, p_2, \ldots, u_n, v_n, p_n$

Figure 7: Block elimination

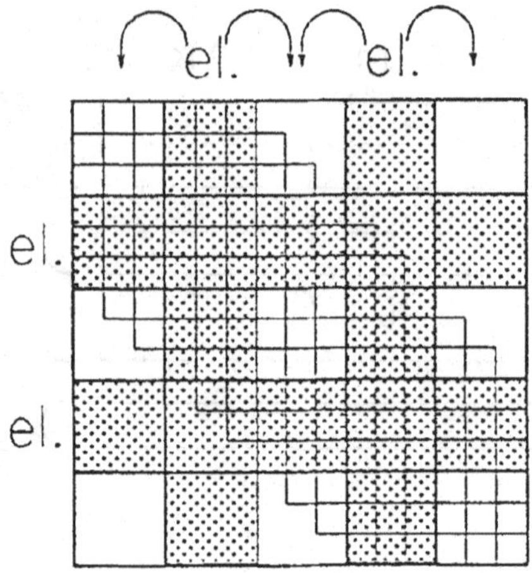

as a single vector of unknowns on the grid a block elimination is performed (see Fig. 7). Fig. 7 now corresponds to a system of 3 PDEs on a 1-D grid with 5 grid points.

9 Implementation

As the matrices have the structure of scattered diagonals they are stored sequentially by the nonzero diagonals. Optimally the diagonals themselves are packed. For the sake of excellent vectorization the matrix-vector multiplication Ax=b which is the governing operation for the PRES methods is formulated by diagonals [8].

Every diagonal d_i is multiplied by the corresponding part of the vector x and the result is added to the appropriate part of the vector b

(21) $$b^{(s)} = b^{(s)} + d_i * x^{(s)}$$

where s means shifted part. In Fig. 8 different geometric symbols indicate which diagonal is multiplied by which part of the vector x and to which part of b the result is added.

Of course this storage scheme of the matrices complicates the elimination process regarding the end rule for an arbitrary number of grid points, the block elimination for 2-D and 3-D problems and the block elimination for systems of partial differential equations. There had to be developed a quite elaborate program, which is further complicated for packed diagonals.

10 Strategies

On each DARE level we have to decide which PRES method has to be applied and how many iterations have to be performed.

Figure 8: Matrix-vector multiplication by diagonals

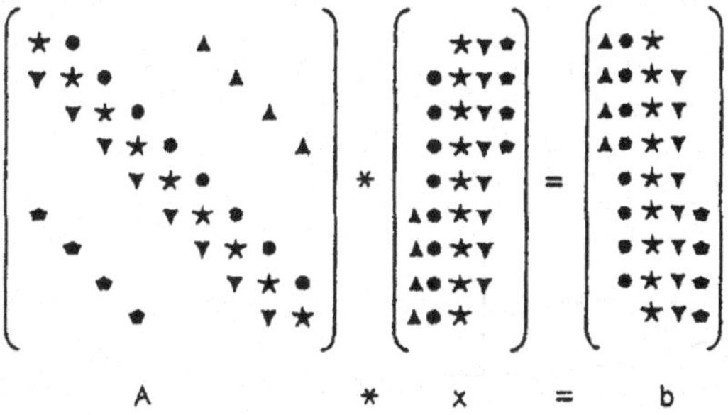

10.1 Choice of algorithm

As there is no iterative solution method which is optimal for all kinds of matrices we developed a polyalgorithm. The PRES20 method is applied to the original system (leading PRES). For the preconditioning on the lowest level the direct Gauss algorithm is used.

On intermediate levels we start with the PRES20 method and perform in any case 3 iteration steps. From the fourth step on the average convergence rate is tested for the iteration step k by the following criterium

$$c = \frac{\sum_{i=1}^{3} ||r_{k+1-i}||}{\sum_{i=1}^{3} ||r_{k-i}||} < \epsilon \tag{22}$$

where r_k is the residual $r_k = Ax_k - b$ for the approximate x_k. If $c \geq \epsilon$ then we consider the iteration as too slow convergent and switch to the ATPRES method. The optimized value for ϵ is at the moment 0.98 (see Fig. 9).

For the ATPRES method the criterium (22) is used with $\epsilon = 0.997$. If c exceeds ϵ the method has "fallen asleep" and further iterations are not usefull. Therefore we switch to the next higher level.

We also tested a three method polyalgorithm, which started with PRES20, continued with the Biconjugate Gradient method and ended with ATPRES. But the Biconjugate Gradient steps didn't improve the convergence, even worse, they only consumed additional time.

10.2 Determination of the number of iterations

On coarse grids iterations are cheap while on fine grids the expensive iterations have to be minimized. This economic consideration suggests to perform many iterations on coarse grids and a few iterations on fine grids. But it makes no sense to produce a very accurate solution on a coarse grid which is only used to interpolate the starting guess for the next higher level. If one doesn't perform iterations on fine grids the information of the lower levels may be lost. The problem to search for an efficient strategy only can be solved by numerical experiments. There are two possibilities to

Figure 9: Polyalgorithm

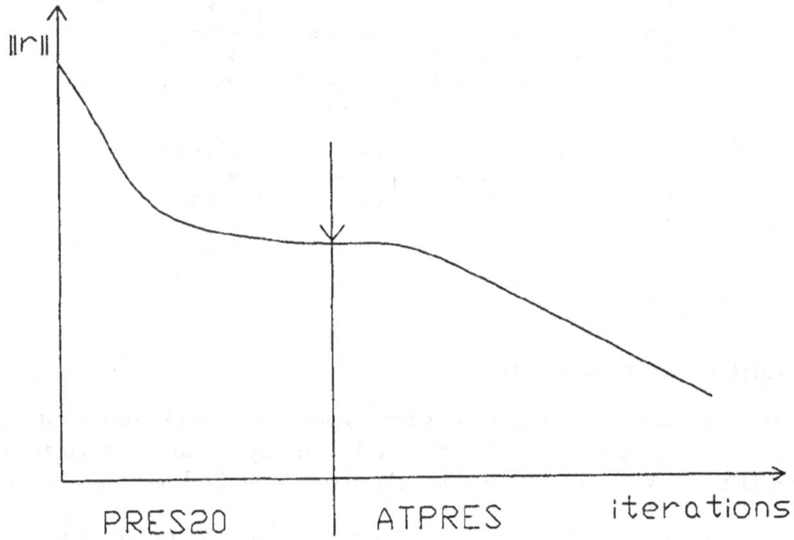

control the procedure

1. a fixed number of steps on each level,
2. a prescribed accuracy on each level.

For well-conditioned matrices resulting for example from the Poisson equation or from fluid dynamic calculations with a low Reynoldsnumber the strategy doesn't matter. If one performs many iterations on one level few cycles are needed. If one performs few iterations on each level more cycles are necessary. The total time is approximately the same. For "well-conditioned" matrices the DARE method is very robust.

The matrices resulting from the discretization and linearization of Navier-Stokes type equations with central difference formulas are for a high Reynoldsnumber ill-conditioned. And for these matrices the optimal number of iterations is dependent on the Reynoldsnumber and the grid size.

10.3 Smoothness control

To overcome the difficulties of the DARE control with ill-conditioned matrices we have to understand when DARE works well and when it doesn't. As it was explained in the section "information transfer" the DARE method works well if the solution is smooth compared to the grid size. The linear systems arise from the linearization and discretization of elliptic and parabolic partial differential equations. The solution of physical problems normally should be smooth. But as the differential equations are nonlinear the Newton method is applied. For this reason the linear equation for the Newton correction is solved. But the solution of this equation has not to be smooth, if the linear systems of the preceding correction steps were solved with a certain accuracy. To guarantee smoothness for the second and the following Newton steps the solution of the linear system has to be performed with the same accuracy requirement as for whole problem solution.

Furthermore in one single Newton step on each level we perform iterations of the appropriate PRES method. But as it is well known these methods may reduce the residual sufficiently while some components of the solution are close to the exact solution and while some components still are far away. Therefore the approximated solution may oscillate which leads to a bad convergence of the DARE method because DARE assumes a smooth solution.

The solution of this problem is apparent. You have to iterate on each level as long as the solution is "rugged". The following control of the smoothness is implemented.

Fit the solution on the index grid by a cubic approximation polynomial P using the five index points k-2,k-1,k,k+1,k+2. Calculate

$$m_k = \max_{k-2 \le i \le k+2} |x_i - P(i)| \,. \tag{23}$$

If the condition

$$m_k < \delta \tag{24}$$

holds for all k then the solution is considered to be smooth and the iteration on this level is terminated. (See Fig. 10). For δ we choose at the moment $\delta = 10^{-1}\eta$ where η is the desired relative reduction for the residual of the original system.

Of course similar to the reduction and interpolation technique for blocks the determination of the smoothness has to be done separately for different directions and different physical components.

This technique established a very robust DARE method which is very close to the optimal "hand control" even for ill-conditioned matrices, and sometimes it is better.

But as all control parameters are closely interconnected we had to change the values of ϵ in equation (22) to the present values 0.995 for PRES20 and 0.995 for ATPRES. There still remains a lot of work to tune the different factors.

10.4 Skipping levels

If the starting residual for a specific level is immediately below the desired accuracy and if the solution is smooth the iteration on this level is skipped.

A further speedup for some cases resulted from a forced skipping the iteration on each second level. As we got the same results from this strategy for the other cases we presently skip each second level (see Fig. 11).

11 Numerical results

All timings are timings for the problem solution, i.e. they include the computation of the difference formulae, the computation of the matrix elements from the Jacobian

Figure 10: Smoothness control

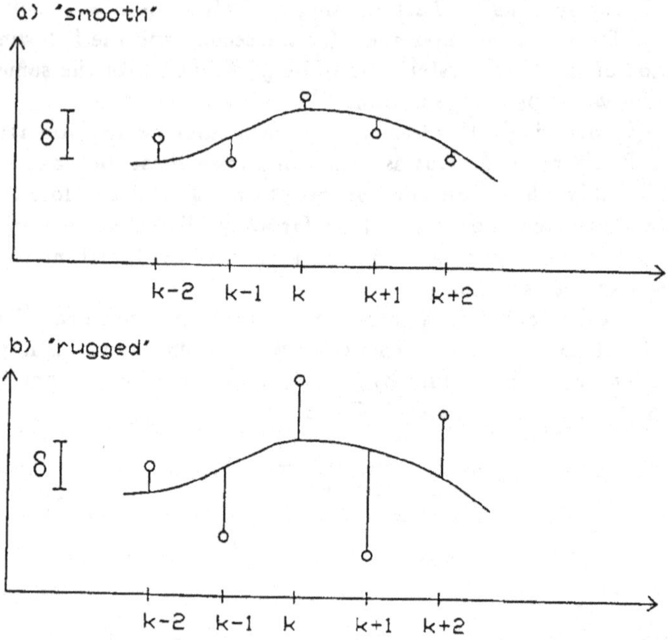

Figure 11: Skipping levels

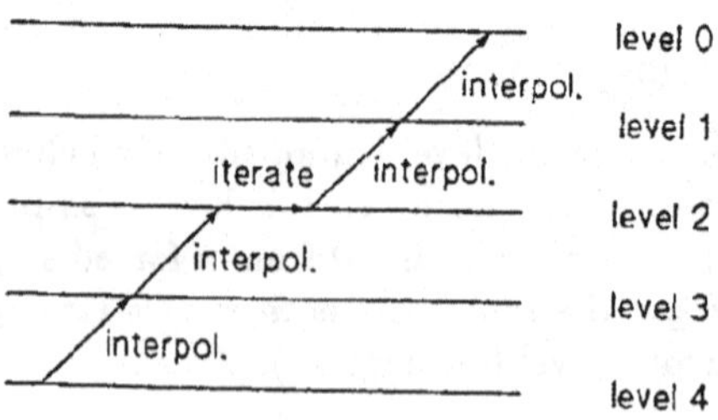

Table 1: Timings for the Poisson equation

grid	unknowns	time for PRES20	time for DARE	speedup	solved systems	DARE cycles
67^2	4489	0.37 s	0.71 s	0.52	4	1/1/2/1
131^2	17161	2.54 s	1.76 s	1.44	4	1/2/2/1
259^2	67081	24.47 s	6.05 s	4.07	5	2/3/3/4/1
515^2	256225	439.49 s	23.32 s	18.84	4	4/3/5/3

Table 2: Increase of unknowns and computation for the Poisson equation compared with the next coarser grid

grid	unknowns	factor of unknowns	factor of time
67^2	4489		
131^2	17161	3.82	2.47
259^2	67081	3.90	3.43
515^2	256225	3.95	3.85

matrices and from higher order difference formulae, they include several Newton steps and the computation of the error estimate of the FIDISOL program, see section 17 in [3]. Thus several linear systems are solved. The measurements were performed on the Siemens (Fujitsu) VP400-EX vector computer at Karlsruhe University.

The first example is the following Poisson equation

(25) $$u_{xx} + u_{yy} + 8\pi^2 sin2\pi x\ cos2\pi y = 0$$

discretized with consistency order 2 on the unit square with Dirichlet boundary conditions. The exact solution is

$$u(x,y) = sin2\pi x\ cos2\pi y\ .$$

The maximum global relative error $\frac{||\Delta u||}{||u||}$ of the computed solution is 10^{-4} in the Euclidean norm. In Table 1 the speedup by DARE preconditioning is listed. The number of Newton steps including the error estimate is listed in the column "solved systems". The number of DARE cycles for the solution of each system is given in the next column. The values in Table 2 show that one gets a nearly linear increase in computation with the number of unknowns. The DARE strategy was a "hand control" with 20 fixed iterations on each level.

The second example is the following nonlinear system of equations

(26) $$\begin{aligned} u_{xx} + u_{yy} + R(uu_x + vu_y) + u - g_1(x,y) = 0\ , \\ v_{xx} + v_{yy} + R(uv_x + vv_y) + v - g_2(x,y) = 0 \end{aligned}$$

with R=1. The system was discretized with consistency order 4 on the unit square with Dirichlet boundary conditions. The exact solution is

$$u(x,y) = sin2\pi x\ cos2\pi y\ ,$$

Table 3: Timing results for the system (26)

grid	unknowns	time for PRES20	time for DARE (hc)	speedup (hc)	time for DARE (sc)	speedup (sc)
65^2	8450	1.35 s	1.41 s	0.95	1.34 s	1.01
129^2	33282	7.56 s	3.64 s	2.07	3.25 s	2.32
257^2	132098	53.17 s	9.83 s	5.41	13.68 s	3.89

Table 4: Increase of unknowns and computation for system (26) compared with the next coarser grid

grid	unknowns	factor of unknowns	factor of time (hc)	factor of time (sc)
65^2	8450			
129^2	33282	3.93	2.58	2.42
257^2	132098	3.96	2.70	4.20

$$v(x, y) = cos2\pi x \; sin2\pi y \; .$$

In Table 3 we note the timings for "hand control" (hc) with 20 iterations on each level and the timings for the smoothness control (sc). The timings indicate that for a large number of grid points the smoothness control still can be improved. The data in Table 4 show also in this case at least a linear increase in computation in dependence of the number of grid points, and what is more a superlinear dependence is achieved by an optimal strategy.

12 Conclusion

DARE turned out to be a feasible and promising technique for the preconditioning of generalized CG methods. The technique shows to be robust, i.e. independent from the operator of the partial differential equation and the consistency order. Similar to the multigrid methods the computation increases linearly with the number of unknowns. As the method is still in the development phase a lot of investigations have to be performed to improve the strategies.

It may turn out that there is no single strategy for all cases. Then an "intelligent" algorithm has to be designed which consists of several strategies and which decides itself which strategy has to be applied. The result would be a self adapting polyalgorithm.

Further investigations will be performed for 3-D problems. It may turn out that an area or spatial reduction and interpolation scheme is superior to the present 1-D alternating direction technique.

In any case we have to answer many questions, not to forget the question for a theory for DARE.

References

[1] W. Hackbusch, Multi-Grid Methods and Applications, Springer Verlag, Berlin, 1985.

[2] L. A. Hageman, D. M. Young, Applied Iterative Methods, Academic Press, New York, London, Toronto, Sydney, San Francisco, 1981.

[3] W. Schönauer, Scientific Computing on Vector Computers, North Holland, Amsterdam, New York, Oxford, Tokyo, 1987.

[4] W. Schönauer, H. Müller, E. Schnepf, Pseudo-Residual Type Methods for the Iterative Solution of Large Linear Systems on Vector Computers, Parallel Computing 85, Elsevier Science Publishers B. V. (North Holland), 1986.

[5] W. Schönauer, M. Schlichte, R. Weiss, Wrong Ways and Promising Ways towards Robust and Efficient Iterative Linear Solvers, Advances in Computer Methods for Partial Differential Equations VI, Publ. IMACS, 1987, pp. 7-14.

[6] W. Schönauer, M. Schlichte, R. Weiss, Numerical Experiments with Data Reduction (DARE) Methods for the Solution of Large Linear Systems on Vector Computers, IMACS 1988, 12th World Congress on Scientific Computation, Proceedings, Paris, edited by R. Vichnevetsky, P. Borne, J. Vignes, 1988.

[7] W. Schönauer, R. Weiss, M. Schlichte, The basic Ideas of the Data Reduction (DARE) Method for the Solution of Large Linear Systems on Vector Computers, ZAMM 69, 1989, pp. 184-185.

[8] R. Weiss, H. Häfner, W. Schönauer, Tuning the Matrix-Vector Multiplication in Diagonal Form, Proceedings of Parallel Computing '89, Leiden, edited by G. R. Joubert, D. J. Evans, F. J. Peters, Elsevier Science Publishers, to appear.

Analysis of a recursive 5-point/9-point factorization method

O.Axelsson and V.Eijkhout
Department of Mathematics
University of Nijmegen, the Netherlands

Abstract
Nested recursive two-level factorization methods for nine-point difference matrices are analyzed. Somewhat similar in construction to multilevel methods for finite element matrices, these methods use recursive red-black orderings of the meshes, approximating the nine-point stencils by five-point ones in the red points and then forming the reduced system explicitly. As this Schur complement is again a nine-point matrix (on a skew grid this time), the process of approximating and factorizing can be applied anew.

Progressing until a sufficiently coarse grid has been reached, this gives a multilevel preconditioner for the original matrix. Solving the levels in V-cycle order will not give an optimal order method, but we show that using certain combinations of V-cycles and W-cycles will give methods of both optimal order of numbers of iterations and computational complexity.

KEYWORDS Algebraic multilevel, Chebyshev polynomial approximation, nine-point differences, optimal order preconditioners

1 Introduction

Recently the first author and P.S. Vassilevski [5, 6] have derived and analyzed algebraic multilevel iteration methods for finite element matrices, in particular for piecewise linear approximations. In the present paper we propose somewhat similar methods for nine-point difference matrices. As nine-point difference approximations can have a higher degree of approximation, while being just as easy to construct as five-point approximations, it is of interest to have preconditioners based directly on the nine-point matrix. Actually, the method proposed in this paper is also applicable to five-point difference matrices.

The methods presented here differ from the methods proposed earlier in the fact that, apart from the top level, they do not use coefficient matrices derived from the differential operator. Instead the Schur complement system of the previous level is used. This is possible as the factorization proposed ensures recursively a nine-point structure on each level. If, namely, a red-black ordering is imposed on the grid, and the nine-point stencil in the red points is suitably modified to a five-point one, elimination of these red points will give a skew nine-point stencil on the black points. A fundamental property of this procedure is that the red points become mutually uncoupled, thus making the solution of systems on each level completely parallel, and enabling the explicit formation of the Schur complement.

Continuing this recursive factorization process until a sufficiently coarse grid is obtained (this grid can be arbitrarily coarse, and for systems on it one can for instance use a direct solution method), we obtain a multilevel preconditioner which can be used in a conjugate gradient or other iterative method. However, using the preconditioner as a simple V-cycle will not give spectral equivalence to the original matrix (we show that the relative condition number grows with the meshsize h as $O(h^{-q})$ for some positive q), so we use the nested polynomial approximations proposed earlier, thereby making the preconditioner into a general W-cycle.

More precisely we show that for polynomial degree $\nu = 2$ the relative condition number is $O(1)$, but the number of arithmetic operations grows linearly with the number of levels, thus giving a method with computational complexity of order $O(h^{-2} \log h^{-1})$. Using alternately first and second (or even third) degree polynomials we show that a method of optimal order of computational complexity results.

Earlier use of similar red-black orderings for five-point difference matrices can be found in [4]. The use of 'intermediate' (i.e., skewed) grids has been described in [12] and in [7] and [10]. A method similar to the present one, but for finite element meshes using bisections of triangles, has been analysed in [2]. A first description of 5-point/9-point methods, but without analysis of the influence of polynomials on the condition number, and not based on using the Schur complement explicitly, can be found in [1].

2 Construction of the level k coefficient and preconditioning matrices

The nested recursive factorisation method resembles in some respects recursive two-level methods proposed in [5, 6]. It differs in the fact that, whereas the two-level methods use a finite element stiffness matrix on every level, the nested recursive factorisation uses on each level the Schur-complement of the previous (finer) level to generate the coefficient matrix on that level.

The nested recursive factorisation can be formally described as follows. Let

$$A^{(k+1)} = \begin{pmatrix} D & C^t \\ C & E \end{pmatrix}$$

(where D and E are positive definite, E is symmetric and D is diagonal) be the coefficient matrix on level $k+1$. The coefficient matrix $A^{(k)}$ on the next level is formed as an approximation of the Schur complement

$$S^{(k)} = E - CD^{-1}C^t$$

in such a way that for all vectors u there exist a $\mu_k \in (0,1)$ such that

$$0 < \mu_k u^t S^{(k)} u \leq u^t A^{(k)} u \leq u^t S^{(k)} u. \tag{1}$$

The exact mechanics of this approximation are discussed in the next section.

We precondition $A^{(k+1)}$ or $S^{(k+1)}$ by

$$M^{(k+1)} = \begin{pmatrix} D & 0 \\ C & Z^{(k)} \end{pmatrix} \begin{pmatrix} I & D^{-1}C^t \\ 0 & I \end{pmatrix} \tag{2}$$

where I stands generically for an identity matrix of the proper order, and $Z^{(k)}$ is an approximation to the Schur complement $S^{(k)}$, defined by

$$\begin{aligned} Z^{(k)^{-1}} &= (I - P_\nu(M^{(k)^{-1}} S^{(k)}))S^{(k)^{-1}}, \quad version(i) \\ Z^{(k)^{-1}} &= (I - P_\nu(M^{(k)^{-1}} A^{(k)}))A^{(k)^{-1}}, \quad version(ii). \end{aligned} \tag{3}$$

Here $P_\nu = P_{\nu_k}^{(k)}$ is a polynomial of degree $\nu = \nu_k$ where the order zero term has coefficient 1. We will specify the choice of the polynomials in section 6.

Note that, if we aim at preconditioned solution of a nine-point system, the matrix on the finest mesh is an $S^{(k+1)}$-matrix. Alternately, one could start out with a mixed five/nine-point system. The matrix on the finest mesh is then an $A^{(k+1)}$-matrix.

3 Recursive derivation of the level k coefficients

In the previous section we left unspecified how the approximation of the Schur complement system $S^{(k)}$ by $A^{(k)}$ is performed. This will be discussed now.

We may consider on each level a nine-point matrix $S^{(k)}$ to be given. A standard red-black structure is then imposed on the grid, and the nine-point stencils are modified to five-point stencils in the red points; they are left untouched in the black points. This uncouples the red points. In the transition from this level to the next the red points are eliminated; as we will show below, in the resulting Schur complement system the black points are then connected by a nine-point stencil but on a skew grid.

The coefficient matrix on this skew grid, i.e., the grid of the black points, is again derived by imposing a red-black structure on these points, and modifying the stencils in the new red points in order to obtain five-point connections. Eliminating the red points will then give nine-point stencils on a horizontal/vertical grid, which is the same structure that we had at the outset, but with different coefficients, and on a double distance grid.

Note that for a nine-point stencil the grid need not be rectangular: the desired structure is also present in meshes that were generated by the 'newest-vertex' method for grid refinement [11]. For such irregular grids the model analysis in this section does not hold in general.

Let us then start off with mixed 5-point/9-point stencils on a red/black grid. These mixed stencils give rise to the $A^{(k)}$ matrices. Suppose that we have a nine-point molecule in the black points, composed of s times $(-\Delta_5^{(h,+)})$ plus t times $(-\Delta_5^{(h,\times)})$, where s and t are positive numbers and

$$-\Delta_5^{(h,+)} = \left(\frac{1}{h^2}\right) \begin{matrix} \cdot & -1 & \cdot \\ -1 & 4 & -1 \\ \cdot & -1 & \cdot \end{matrix}$$

$$-\Delta_5^{(h,\times)} = \left(\frac{1}{(\sqrt{2}h)^2}\right) \begin{matrix} -1 & \cdot & -1 \\ \cdot & 4 & \cdot \\ -1 & \cdot & -1 \end{matrix}$$

with grid spacing h in both stencils are the axi-parallel and skew five-point difference molecules, i.e., our initial molecule in the black points is

$$\left(\frac{1}{h^2}\right) \begin{matrix} -t/2 & -s & -t/2 \\ -s & 4s+2t & -s \\ -t/2 & -s & -t/2 \end{matrix} .$$

The stencil in the red points is then found by simply eliminating some connections of the stencil in the black points in order to uncouple the red points, ie., the stencil is

$$\begin{matrix} \cdot & -s & \cdot \\ -s & 4s & -s \\ \cdot & -s & \cdot \end{matrix}$$

Now, in order to eliminate the red points, i.e., the horizontal and vertical neighbours of the black points, we multiply on the horizontal and vertical crossconnections by

$$\begin{matrix} \cdot & 1/4 & \cdot \\ 1/4 & 1 & 1/4 \\ \cdot & 1/4 & \cdot \end{matrix}$$

This gives a resulting (skewed) nine-point stencil

$$\left(\frac{1}{2h^2}\right) \begin{matrix} \cdot & \cdot & -s/2 & \cdot & \cdot \\ \cdot & -(s+t) & \cdot & -(s+t) & \cdot \\ -s/2 & \cdot & 6s+4t & \cdot & -s/2 \\ \cdot & -(s+t) & \cdot & -(s+t) & \cdot \\ \cdot & \cdot & -s/2 & \cdot & \cdot \end{matrix}$$

(or $(s+t)(-\Delta_5^{(h,\times)}) + t(-\Delta_5^{(2h,+)})$) on the black points, corresponding to the Schur complement matrix $S^{(k-1)}$.

Next we impose a red-black structure on the black points and modify the stencil in the red points by eliminating the horizontal/vertical connections and moving them to the central node, i.e., we eliminate the $s(-\Delta_5^{(2h,+)})$ term at the red points. Thus we obtain the $A^{(k-1)}$ matrix corresponding to the skew grid, which has the above stencil in the black points, and the stencil

$$\left(\frac{1}{2h^2}\right) \begin{matrix} -(s+t) & \cdot & -(s+t) \\ \cdot & 4(s+t) & \cdot \\ -(s+t) & \cdot & -(s+t) \end{matrix}$$

in the red points. Eliminating the red points of the skew grid amounts to multiplying by the stencil

$$\begin{matrix} 1/4 & \cdot & 1/4 \\ \cdot & 1 & \cdot \\ 1/4 & \cdot & 1/4 \end{matrix}$$

on the skew connections of the black points. This gives the Schur complement system $S^{(k-2)}$ with the corresponding double distance nine-point molecule

$$\left(\frac{1}{2h}\right)^2 \begin{matrix} -(t+s)/2 & \cdot & -(2s+t) & \cdot & -(s+t)/2 \\ \cdot & \cdot & \cdot & \cdot & \cdot \\ -(2s+t) & \cdot & 10s+6t & \cdot & -(2s+t) \\ \cdot & \cdot & \cdot & \cdot & \cdot \\ -(t+s)/2 & \cdot & -(2s+t) & \cdot & -(s+t)/2 \end{matrix}$$

or $(2s+t)(-\Delta_5^{(2h,+)})+(s+t)(-\Delta_5^{(2h,\times)})$. Thus, as is readily seen, the factorization progresses with stencils

$$\begin{array}{lll} \text{on level} & k & a_k(-\Delta_5^{(h,+)}) \\ & & b_k(-\Delta_5^{(h,\times)}) \\ & k-1 & a_{k-1}(-\Delta_5^{(h,\times)}) \\ & & b_{k-1}(-\Delta_5^{(2h,+)}) \\ & k-2 & a_{k-2}(-\Delta_5^{(2h,+)}) \\ & & b_{k-2}(-\Delta_5^{(2h,\times)}) \end{array}$$

and coefficients satisfying

$$a_{k-1} = a_k + b_k, \quad b_{k-1} = a_k \qquad \text{for } k = p, p-1, \ldots, 1.$$

We see that (after a normalization) the coefficients satisfy a Fibbonacci series, with the coefficients for the finest mesh (level p) given. In explicit form

$$a_k = c_1\left(\frac{1+\sqrt{5}}{2}\right)^{p-k} + c_2\left(\frac{1-\sqrt{5}}{2}\right)^{p-k}$$

with

$$c_1 = \frac{1}{2}(1+\frac{1}{\sqrt{5}})a_p + \frac{1}{\sqrt{5}}b_p, \qquad c_2 = \frac{1}{2}(1-\frac{1}{\sqrt{5}})a_p - \frac{1}{\sqrt{5}}b_p.$$

In the course of the factorization (that is, for $k \downarrow 0$) the quotient b_k/a_k converges as

$$\frac{b_k}{a_k} = \frac{a_{k+1}}{a_k} \to \sigma = \frac{\sqrt{5}-1}{2} \approx 0.618;$$

its maximal value is $a_p/(a_p+b_p)$ if $c_2 > 0$, it is b_p/a_p if $c_2 < 0$. Furthermore, if $b_p = \sigma a_p$ then $b_k/a_k = \sigma$ for all k.

4 Local analysis

In this section we analyse the error that we make by modifying a Schur complement system $S^{(k)}$ to obtain the 5-point/9-point coefficient matrix $A^{(k)}$ on that level. As was seen in the previous section, the Schur complement matrix itself was derived by eliminating the red points in the coefficient matrix $A^{(k+1)}$ one level higher.

For the approximation analysis we construct element matrices that can be associated with the global matrices. We analyse then the relative condition numbers of the element matrices associated with the $A(k)$ and $S(k)$ matrices. We give this analysis only for isotropic problems, the anisotropic case will be treated more fully elsewhere.

Now suppose that on a certain (horizontal/vertical) level we have a coefficient matrix $A^{(k+1)}$ in which the black points are equiped with a nine-point stencil

$$\begin{matrix} -b/2 & -a & -b/2 \\ -a & 4a+2b & -a \\ -b/2 & -a & -b/2 \end{matrix}$$

then the Schur complement matrix $S^{(k)}$ obtained after eliminating the red points will have associated with it the nine-point stencil

$$\begin{matrix} . & . & -a/4 & . & . \\ . & -(b+a)/2 & . & -(b+a)/2 & . \\ -a/4 & . & 3a+2b & . & -a/4 \\ . & -(b+a)/2 & . & -(b+a)/2 & . \\ . & . & -a/4 & . & . \end{matrix}$$

We can construct local element matrices corresponding to the above Schur complement, without any need to actually construct any underlying basisfunctions. This process is simply the reverse of the usual course of action when a global matrix is assembled from local element matrices.

Consider then a local element

$$\begin{matrix} & (1) & \\ (3) & & (4) \\ & (2) & \end{matrix}$$

on the skew grid of level k (these are the the black points of level $k+1$). Each center node has support of four skewed squares, and we find the element matrix

$$S_e = \frac{1}{4}\begin{pmatrix} (3a+2b) & -a & -(b+a) & -(b+a) \\ -a & (3a+2b) & -(b+a) & -(b+a) \\ -(b+a) & -(b+a) & (3a+2b) & -a \\ -(b+a) & -(b+a) & -a & (3a+2b) \end{pmatrix}$$

Letting the points (3) and (4) be the red points of the next level, we derive the matrix $A^{(k)}$ by imposing on these red points a five-point stencil

$$\begin{matrix} -(b+a)/2 & . & -(b+a)/2 \\ . & 2a+2b & . \\ -(b+a)/2 & . & -(b+a)/2 \end{matrix}$$

found by eliminating the couplings between the red points. For the coefficient matrix $A^{(k)}$ we then find an element matrix

$$A_e = \frac{1}{4}\begin{pmatrix} (3a+2b) & -a & -(b+a) & -(b+a) \\ -a & (3a+2b) & -(b+a) & -(b+a) \\ -(b+a) & -(b+a) & (2a+2b) & 0 \\ -(b+a) & -(b+a) & 0 & (2a+2b) \end{pmatrix}$$

Note that the (2,2) block of this matrix is diagonal, corresponding to the fact that nodes (3) and (4) of the local element are red nodes, which are uncoupled.

Now we want to solve the generalized eigenvalue problem

$$S_e \begin{pmatrix} x \\ y \end{pmatrix} = \lambda A_e \begin{pmatrix} x \\ y \end{pmatrix}.$$

Since S_e and A_e have the same nullspace, the eigenvalue problem is understood as finding the solutions in the complementary space to the nullspace. For this we partition the matrices into 2×2 blocks and make a block factorization

$$S_e = \begin{pmatrix} S_{11} & S_{12} \\ S_{21} & S_{22} \end{pmatrix} = \begin{pmatrix} S_{11} & 0 \\ S_{21} & I \end{pmatrix} \begin{pmatrix} I & S_{11}^{-1} S_{12} \\ 0 & S_{22} - S_{21} S_{11}^{-1} S_{12} \end{pmatrix},$$

$$A_e = \begin{pmatrix} S_{11} & S_{12} \\ S_{21} & A_{22} \end{pmatrix} = \begin{pmatrix} S_{11} & 0 \\ S_{21} & I \end{pmatrix} \begin{pmatrix} I & S_{11}^{-1} S_{12} \\ 0 & A_{22} - S_{21} S_{11}^{-1} S_{12} \end{pmatrix}$$

and we find that the solution to the generalized eigenvalue problem satisfies:

1. $\lambda = 1$ and $S_{22} y = A_{22} y$, or
2. $x = -S_{11}^{-1} S_{12} y$ and $(S_{22} - S_{21} S_{11}^{-1} S_{12}) y = \lambda (A_{22} - S_{21} S_{11}^{-1} S_{12}) y$, where y is not in the nullspace of the two matrices.

An elementary computation reveals that the vector $y = \binom{1}{1}$ satisfies $S_{22} y = A_{22} y$, so the generalized eigenvalue problem in the second point simplifies to

$$\frac{2a+b}{a+b} \begin{pmatrix} 1 & -1 \\ -1 & 1 \end{pmatrix} \begin{pmatrix} y_1 \\ y_2 \end{pmatrix} = \lambda \begin{pmatrix} 1 & -1 \\ -1 & 1 \end{pmatrix} \begin{pmatrix} y_1 \\ y_2 \end{pmatrix}.$$

Hence, the eigenvalues to the generalized eigenproblem are

$$\lambda = 1 \qquad \text{and} \qquad \lambda = \frac{2a+b}{a+b}, \tag{4}$$

the latter for a vector with $y_1 \neq y_2$.

Changing a nine-point matrix into a five-point by the process of moving elements to the central node corresponds to subtracting a certain positive semi-definite matrix, so from the positiveness of a and b at every level we conclude that the coefficient matrix $A^{(k)}$ and the Schur complement matrix $S^{(k)}$ derived from the previous level satisfy

$$\text{for all } x: \quad \mu_k x^t S^{(k)} x \le x^t A^{(k)} x \le x^t S^{(k)} x \tag{5}$$

where (4) shows that $\mu_k = (a+b)/(2a+b)$, i.e., $\mu_k \in (0.5, 1)$.

Similar to the corresponding analysis in [5, 6], we find that if we construct the nine-point stencil recursively from an initial coarse grid to finer levels, we can even permit the coefficients of the underlying elliptic differential equation to be discontinuous on the elements of the coarsest grid. The quantity μ_k can still be computed locally. The relation between μ_k and the parameter γ_k used in [5, 6] is $\mu_k = 1 - \gamma_k^2$. Thus the method described in this paper bears a relation to methods using a splitting of the finite element space and analysing convergence using the quantity γ_k from the Cauchy-Buniakowsky-Schwarz inequality (see [9]). Another multilevel method that relaxes the conditions of the CBS inequality can be found in [8].

As an example, the above estimates are valid for the nine-point discretization of $-\underline{\nabla} \rho \underline{\nabla} u = f$ in Ω, where Ω consists of four rectangular boxes with $\rho = \rho_i$, $i =$

1, 2, 3, 4.

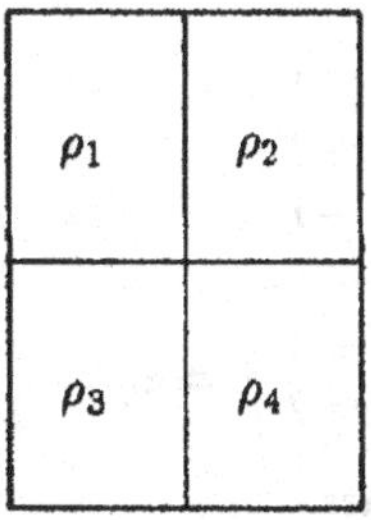

In particular, they are valid for $\rho_4 = \infty$, which corresponds to a problem on an L-shaped (or 're-entrant corner') domain, with Neumann type boundary conditions in the interior boundary part of the 'L'.

Equation 5 could also have been derived in the following way. Consider the original nine-point matrix and its approximation by using a five-point stencil in the red points. From the generalised eigenvalue problem for the corresponding element matrices we find in this case

$$\mu_p = a_p/(a_p + b_p) = a/(a + b).$$

The relation between the entries in the element matrices for the original mesh and the next, skew, mesh is

$$b \leftarrow a/2, \qquad a \leftarrow (b + a)/2.$$

Hence

$$\mu_{p-1} = \frac{(b + a)/2}{(b + a)/2 + a/2} = \frac{b + a}{b + 2a}$$

which gives 5. Similarly, using the recurrence relations

$$a_{k-1} = a_k + b_k, \qquad b_{k-1} = a_k$$

we find

$$\mu_k = \frac{1}{1 + \mu_{k-1}}, \qquad k = p, p - 1, \ldots, 1$$

which shows that the μ_k converge to the positive solution of the equation $\mu^2 + \mu - 1 = 0$, i.e., $\mu = \sigma \approx 0.618$.

We will now calculate the value for the finest mesh (say at level p) of μ_p for some nine-point stencils.

For the standard nine-point difference stencil (which is of fourth order of approximation after suitable modification of the right hand side)

$$\begin{matrix} -1 & -4 & -1 \\ -4 & 20 & -4 \\ -1 & -4 & -1 \end{matrix}$$

we have

$$\left.\begin{matrix} a = 4 \\ b = 2 \end{matrix}\right\} \Rightarrow \mu_p = \frac{4}{4 + 2} = \frac{2}{3}$$

which is the largest μ_k in this case. For the nine-point molecule arising from bilinear finite elements

$$\begin{array}{ccc} -1 & -1 & -1 \\ -1 & 8 & -1 \\ -1 & -1 & -1 \end{array}$$

we find

$$\left.\begin{array}{l} a = 1 \\ b = 2 \end{array}\right\} \Rightarrow \mu_p = \frac{1}{3},$$

which is the smallest μ_k in this case.

5 Computational complexity

The solution of a system $M^{(k+1)}a = b$ involves the solution of a system of lower dimension with $Z^{(k)}$. From the definition of $Z^{(k)}$, (3), we see that solving

$$Z^{(k)}y = x$$

is equivalent to computing

$$y = \left(I - P_\nu({M^{(k)}}^{-1} S^{(k)})\right){S^{(k)}}^{-1}x \qquad \text{for version (i)} \tag{6}$$

or

$$y = \left(I - P_\nu({M^{(k)}}^{-1} A^{(k)})\right){A^{(k)}}^{-1}x \qquad \text{for version (ii).} \tag{7}$$

As was already remarked before, we choose for P_ν a polynomial which can be written

$$P_\nu(x) = 1 - c_1x - \cdots - c_\nu x^\nu,$$

so for instance the solution process (6) boils down to

$$\begin{aligned} y &= \left[c_\nu\left({M^{(k)}}^{-1}S^{(k)}\right)^{\nu-1}{M^{(k)}}^{-1} + \cdots + c_1{M^{(k)}}^{-1}\right]x \\ &= {M^{(k)}}^{-1}\left[c_1 + S^{(k)}{M^{(k)}}^{-1}(\cdots(c_{\nu-1} + S^{(k)}{M^{(k)}}^{-1}c_\nu)\cdots)\right]x, \end{aligned}$$

which can be written algorithmically as

$$\begin{array}{l} y \leftarrow 0 \\ \text{for } k = 0, \ldots, \nu - 1 \text{ do} \\ \quad y \leftarrow {M^{(k)}}^{-1}\left(c_{\nu-k}x + S^{(k)}y\right) \end{array}$$

for version (i); for version (ii) we must replace $S^{(k)}$ by $A^{(k)}$.

We will now proceed to give operation counts for the solution of a system with preconditioners using combined five-point/nine-point stencils and the elimination procedure described above. If we take N to denote the number of points on the finest mesh, i.e., the mesh at level p, the number of points N_k on the mesh at level k equals (disregarding boundary effects)

$$N_k = N/2^{p-k}.$$

From the definition 2 of the preconditioner and the algorithm to solve systems with $Z^{(k)}$ above, we find that the number of operations ω_{k+1} needed to solve a system with $M^{(k+1)}$ can (for version (i)) be given by

$$\begin{aligned}
\omega_{k+1} &= 2 \cdot N_k && \text{solve with } D^{(k)} \text{ twice} \\
&\quad + 2 \cdot 4 \cdot N_k && \text{multiply by } C \text{ and } C^t \\
&\quad + \nu\omega_k + (\nu + 9(\nu - 1)) \cdot N_k && \text{solve system with } Z^{(k)} \\
&= (5\nu + 1/2)N_{k+1} + \nu\omega_k && \text{(Total)}
\end{aligned}$$

For version (ii) we find similarly

$$\omega_{k+1} = (4\nu + 3/2)N_{k+1} + \nu\omega_k.$$

These operation counts concern the number of multiplications. The number of additions is somewhat smaller. When the coefficients are locally constant we can actually reduce the number of multiplications substantially, using a method of summations of differences, or taking averages for the computation of matrix vector multiplications. For further details, see [3]. However, in the interest of not presenting too many variations of our method, we do not give the corresponding work estimate here.

It is now easy to estimate the amount of work for various values of ν. Let the amount of work ω_0 at the coarsest level be given. As $N_k = N_{k+1}/2$ we can solve the recurrence for the amount of work ω_k explicitly:

$$\omega_k = (5\nu + 1/2)(1 + \cdots + (\frac{\nu}{2})^{k-1})N_k + \nu^k\omega_0.$$

For some specific values of ν we find for the total amount of work at the finest level

$$\begin{aligned}
\nu = 1:&\ \omega_p \leq 11N_p + \omega_0 \\
\nu = 2:&\ \omega_p \leq 10.5pN_p + 2^p\omega_0 \\
\nu = 3:&\ \omega_p \leq 31(\tfrac{3}{2})^p N_p + 3^p\omega_0
\end{aligned}$$

for version (i); for version (ii) we find somewhat smaller constants, except for $\nu = 1$ when the methods coincide. Thus already for $\nu = 2$ the preconditioner can no longer be of optimal order of computational complexity.

As it is desirable to have polynomial degrees higher than 1 (see the analysis of the condition number in the next section), we shall propose a preconditioner that uses different polynomials on different levels.

In particular let us consider a preconditioner which uses first and either second or third degree polynomials on alternate levels. Let k be such that a first degree polynomial is used on level $k+1$. For the amount of work we then find for version (i)

$$\begin{aligned}
\omega_{k+2} &= 5.5N_{k+2} + (5\nu + 1/2)N_{k+1} + \nu\omega_k \\
&= (23/4 + 5\nu/2)N_{k+2} + \nu\omega_k
\end{aligned}$$

The explicit solution of this recurrence is

$$\omega_{2k} = (23/4 + 5\nu/2)(1 + \cdots + (\nu/4)^{k-1})N_{2k} + \nu^k\omega_0$$

which gives (neglecting the last term)

$$\nu = 2 : \omega_p \leq 21.5N; \qquad \nu = 3 : \omega_p \leq 53N$$

for version (i), and

$$\nu = 2 : \omega_p \leq 20.5N; \qquad \nu = 3 : \omega_p \leq 49N$$

for version (ii). Hence both versions of the method give, when using on alternating levels polynomials of degree 1 and 2 or 3, a preconditioner with computational work per preconditioning step of optimal order, $O(N)$.

6 Estimation of the condition number

In this section we analyse the relative condition numbers of the coefficient matrix $A^{(k)}$ and the Schur complement matrix $S^{(k)}$ to the preconditioner $M^{(k)}$ on the k-th level. This is done by comparing the condition numbers on two consecutive levels and performing a limit analysis. We show that if the polynomial P_ν in the preconditioner is taken as a linear polynomial, the relative condition number increases with the meshsize h as h^{-q}, where $q = \log_2 \sigma^{-1}, \sigma \approx 0.618$; if a higher degree polynomial is taken the condition number can be reduced to $O(1)$.

However, as per level only half of the points are eliminated, such higher degree polynomials do not give a preconditioner of an optimal order in the number of arithmetic operations. Therefore we also consider preconditioners alternately taking polynomials of first and higher degrees; this will again lead to an $O(1)$ condition number. In the previous section it was already shown that such preconditioners have an optimal order of the number of arithmetic operations for polynomial degrees 2 and 3.

We begin by recalling that the preconditioner $M^{(k+1)}$ satisfies

$$M^{(k+1)} = \begin{pmatrix} D & C^t \\ C & Z^{(k)} + CD^{-1}C^t \end{pmatrix} = A^{(k+1)} + \begin{pmatrix} 0 & 0 \\ 0 & Z^{(k)} - S^{(k)} \end{pmatrix} \tag{8}$$

where

$$S^{(k)} = E - CD^{-1}C^t$$

and where $Z^{(k)}$ is defined as

$$Z^{(k)^{-1}} = (I - P_\nu(M^{(k)^{-1}} S^{(k)}))S^{(k)^{-1}}$$

for version (i), and

$$Z^{(k)^{-1}} = (I - P_\nu(M^{(k)^{-1}} A^{(k)}))A^{(k)^{-1}}$$

for version (ii). At the lowest level $k = 0$ (i.e., the coarsest mesh) we let

$$A^{(0)} = M^{(0)}.$$

The polynomials P_ν can have both their degree and their coefficients defined per level; they are chosen to have

$$\text{for all } x \in I_k: \qquad 0 \leq P(x) < 1,$$

where the interval I_k is defined as

$$\text{for version (i):} \qquad I_k = \left[\inf_u \frac{u^t S^{(k)} u}{u^t M^{(k)} u}, \sup_u \frac{u^t S^{(k)} u}{u^t M^{(k)} u}\right],$$

$$\text{for version (ii):} \qquad I_k = \left[\inf_u \frac{u^t A^{(k)} u}{u^t M^{(k)} u}, \sup_u \frac{u^t A^{(k)} u}{u^t M^{(k)} u}\right].$$

Also we assume that P is chosen to have at least one zero in I_k.

For the further analysis we need the following basic lemma.

Lemma 1 *The following relations between coefficient matrices and Schur complement matrices hold.*

1. *For all* v

$$0 < \mu_k v^t S^{(k)} v \leq v^t A^{(k)} v \leq v^t S^{(k)} v.$$

2. *For all* $v = (v_1, v_2)$ *partitioned consistently with* $A^{(k+1)}$

$$v^t A^{(k+1)} v \geq v_2^t S^{(k)} v_2$$

and equality is attainable.

Proof. Part 1 is formula 1; the value of μ_k was derived in formula 5. Part 2 follows by noting that

$$\begin{aligned} v^t A^{(k+1)} v &= (v_1 - D^{-1}C^t v_2)^t D (v_1 - D^{-1} C^t v_2) \\ &\quad + v_2^t (E - C D^{-1} C^t) v_2 \\ &\geq v_2^t S^{(k)} v_2 \end{aligned}$$

with equality if $v_1 = D^{-1} C^t v_2$. Qed.

The definition of the polynomial implies the following a priori bounds on the relative condition of $S^{(k)}$ and $Z^{(k)}$:

Lemma 2 *For version (i) of the method*

$$\text{for all vectors } u \qquad \frac{u^t S^{(k)} u}{u^t Z^{(k)} u} \in (0, 1]; \tag{9}$$

for version (ii) of the method

$$\text{for all vectors } u \qquad \frac{u^t S^{(k)} u}{u^t Z^{(k)} u} \in (0, \mu_k^{-1}], \tag{10}$$

with the lower and upper bounds satisfying

$$\inf_u \frac{u^t S^{(k)} u}{u^t Z^{(k)} u} \leq 1 \leq \sup_u \frac{u^t S^{(k)} u}{u^t Z^{(k)} u}.$$

Proof. The assertion for version (i) follows trivially from the definition of the polynomial. In order to obtain the bounds for version (ii) use in addition lemma 1, and note that for block vectors with a zero first block component

$$\frac{u^t S^{(k)} u}{u^t Z^{(k)} u} = \frac{u^t A^{(k)} u}{u^t Z^{(k)} u} \in (0, 1],$$

which implies by the definition of the polynomial that the lower bound is indeed ≤ 1. Also, for any $u \neq 0$

$$\frac{u^t S^{(k)} u}{u^t Z^{(k)} u} = \frac{u^t S^{(k)} u}{u^t A^{(k)} u} \frac{u^t A^{(k)} u}{u^t Z^{(k)} u} \leq \mu_k^{-1}.$$

If u is such that $P(u^t A^{(k)} u / u^t M^{(k)} u) = 0$ (and note that we have required P to have a zero in I_k) then we find

$$\frac{u^t S^{(k)} u}{u^t Z^{(k)} u} \geq \frac{u^t A^{(k)} u}{u^t Z^{(k)} u} = 1,$$

which establishes that the upper bound is ≥ 1. Qed.

Investigation of $u^t S^{(k)} u / u^t Z^{(k)} u$ gives us bounds for $u^t A^{(k+1)} u / u^t M^{(k+1)} u$:

Lemma 3 *For both versions of the method*

$$\min_u \frac{u^t A^{(k+1)} u}{u^t M^{(k+1)} u} = \min_{u_2} \frac{u_2^t S^{(k)} u_2}{u_2^t Z^{(k)} u_2},$$

and

$$\textit{for version (i):} \qquad \max_u \frac{u^t A^{(k+1)} u}{u^t M^{(k+1)} u} = 1;$$

$$\textit{for version (ii):} \qquad \max_u \frac{u^t A^{(k+1)} u}{u^t M^{(k+1)} u} = \max_{u_2} \frac{u_2^t S^{(k)} u_2}{u_2^t Z^{(k)} u_2}.$$

Proof. For the lower bound let

$$\tau = \inf_u \frac{u^t S^{(k)} u}{u^t Z^{(k)} u},$$

and note that lemma 2 shows that $\tau \leq 1$; using (8) the bound then follows from

$$\begin{aligned} \frac{u^t M^{(k+1)} u}{u^t A^{(k+1)} u} &= 1 + \frac{u^t (M^{(k+1)} - A^{(k+1)}) u}{u^t A^{(k+1)} u} \\ &= 1 + \frac{u_2^t (Z^{(k)} - S^{(k)}) u_2}{u^t A^{(k+1)} u} \leq 1 + (\tau^{-1} - 1) \frac{u_2^t S^{(k)} u_2}{u^t A^{(k+1)} u} \leq \tau^{-1}, \quad (11) \end{aligned}$$

where the last inequality follows from $\tau \leq 1$ and lemma 1.

For the upper bound we have to consider the two versions of the method separately. For version (i) equations (8) and (9) imply that

$$u^t M^{(k+1)} u = u^t A^{(k+1)} u + u_2^t (Z^{(k)} - S^{(k)}) u_2 \geq u^t A^{(k+1)} u;$$

for version (ii) let

$$\tau = \inf_{u_2} \frac{u_2 Z^{(k)} u_2}{u_2 S^{(k)} u_2},$$

and note that $\tau \leq 1$ by lemma 2. Now we have for all vectors $u = (u_1, u_2)$

$$(1 - \tau) u^t A^{(k+1)} u \geq (1 - \tau) u_2^t S^{(k)} u_2 \geq u_2^t (S^{(k)} - Z^{(k)}) u_2$$

so

$$u^t M^{(k+1)} u = u^t A^{(k+1)} u + u_2^t (Z^{(k)} - S^{(k)}) u_2 \geq \tau u^t A^{(k+1)} u.$$

Also in this case we have equality for the vectors $u = (u_1, u_2)$ for which $S^{(k)} u_2 = A^{(k+1)} u$.
Qed.

Lemma 4 *The following upper bounds hold for the quadratic forms of $A^{(k+1)}$ and $S^{(k+1)}$ relative to $M^{(k+1)}$: for version (i)*

$$\frac{u^t A^{(k+1)} u}{u^t M^{(k+1)} u} \leq 1; \qquad \frac{u^t S^{(k+1)} u}{u^t M^{(k+1)} u} \leq \mu_{k+1}^{-1},$$

and for version (ii)

$$\frac{u^t A^{(k+1)} u}{u^t M^{(k+1)} u} \leq \mu_k^{-1}; \qquad \frac{u^t S^{(k+1)} u}{u^t M^{(k+1)} u} \leq \mu_k^{-1} \mu_{k+1}^{-1}.$$

Proof. We proved the bounds for $A^{(k+1)}$ in the previous lemma. The bounds for $S^{(k+1)}$ then follow from lemma 1. Possibly the bound for version (ii) could be slightly improved by a refined analysis, but we believe that the bound is essentially of the right form.
Qed.

Corollary 1 *The interval I_k can be given by*

$$I_k = \left[\inf_u \frac{u^t S^{(k-1)} u}{u^t Z^{(k-1)} u}, \mu_{k-\sigma}^{-1} \right]. .$$

where $\sigma = 0$ for version (i) and $\sigma = 1$ for version (ii).

Proof. Combine lemma's 3 and 4, and for the lower bound for version (i) use in addition that

$$\frac{u^t S^{(k)} u}{u^t M^{(k)} u} \geq \inf \frac{u^t S^{(k)} u}{u^t A^{(k)} u} \frac{u^t A^{(k)} u}{u^t M^{(k)} u}$$

for all $u \neq 0$, and lemma 1a.
Qed.

We will now give the definition of the polynomials $P_{\nu_k}^{(k)}$. The best approximation to zero on the interval I_k among polynomials of degree ν satisfying $0 \leq P_\nu(x) < 1$ and $P_\nu(0) = 1$, is the shifted and scaled Chebyshev polynomial

$$P_{\nu_k}^{(k)}(t) = \frac{1 + T_{\nu_k}\left(\frac{\beta_k + \alpha_k - 2t}{\beta_k - \alpha_k}\right)}{1 + T_{\nu_k}\left(\frac{\beta_k + \alpha_k}{\beta_k - \alpha_k}\right)} \tag{12}$$

where $T_\nu(x) = \frac{1}{2}[(x + \sqrt{x^2 - 1})^\nu + (x - \sqrt{x^2 - 1})^\nu]$ and

$$\alpha_k = \inf_u \frac{u^t S^{(k-1)} u}{u^t Z^{(k-1)} u}$$

$$\beta_k = \begin{cases} \mu_k^{-1} & \text{version (i)} \\ \mu_{k-1}^{-1} & \text{version (ii)} \end{cases}$$

Remark 1 *For $\nu_k = 1$ the polynomial reduces to*

$$P_{\nu_k}^{(k)}(t) = 1 - t/\beta_k = \begin{cases} 1 - \mu_k t \\ 1 - \mu_{k-1} t \end{cases}$$

for versions (i) and (ii) respectively. *Qed.*

Lemma 5 *If we define ϕ_k by*

for version (i): $\phi_k = \inf_u \frac{u^t S^{(k)} u}{u^t M^{(k)} u}$,

for version (ii): $\phi_k = \inf_u \frac{u^t A^{(k)} u}{u^t M^{(k)} u}$,

then a lower bound for $u^t S^{(k)} u / u^t Z^{(k)} u$ is given by

$$\frac{u^t S^{(k)} u}{u^t Z^{(k)} u} \geq 1 - P_\nu^{(k)}(\phi_k)$$

for both versions.

Proof. The assertion follows as

$$\frac{u^t S^{(k)} u}{u^t Z^{(k)} u} = 1 - P_{\nu_k}^{(k)}\left(\frac{\tilde{u}^t S^{(k)} \tilde{u}}{\tilde{u}^t M^{(k)} \tilde{u}}\right) \qquad \text{for some } \tilde{u}$$

for version (i), and

$$\frac{u^t S^{(k)} u}{u^t Z^{(k)} u} = \frac{u^t S^{(k)} u}{u^t A^{(k)} u}\left(1 - P_{\nu_k}^{(k)}\left(\frac{\tilde{u}^t A^{(k)} \tilde{u}}{\tilde{u}^t M^{(k)} \tilde{u}}\right)\right) \qquad \text{for some } \tilde{u}$$

for version (ii), where also lemma 1 has been used. Qed.

Remark 2 *Note that also for version (i) lemma 3 shows that*

$$\text{for all } u: \qquad \frac{u^t A^{(k+1)} u}{u^t M^{(k+1)} u} \geq \phi_k.$$

Qed.

We collect the above results into a theorem.

Theorem 1 *The following bounds hold for the quadratic forms of $A^{(k+1)}$ and $S^{(k+1)}$ relative to $M^{(k+1)}$: for version (i)*

$$\frac{u^t A^{(k+1)} u}{u^t M^{(k+1)} u} \in [1 - P_{\nu_k}^{(k)}(\phi_k), 1]; \qquad \frac{u^t S^{(k+1)} u}{u^t M^{(k+1)} u} \in [1 - P_{\nu_k}^{(k)}(\phi_k), \mu_{k+1}^{-1}],$$

and for version (ii)

$$\frac{u^tA^{(k+1)}u}{u^tM^{(k+1)}u} \in [1 - P^{(k)}_{\nu_k}(\phi_k), \mu_k^{-1}]; \qquad \frac{u^tS^{(k+1)}u}{u^tM^{(k+1)}u} \in [1 - P^{(k)}_{\nu_k}(\phi_k), \mu_k^{-1}\mu_{k+1}^{-1}],$$

where the quantities ϕ_k satisfy the recurrence

$$\phi_k = 1 - P^{(k-1)}_{\nu_{k-1}}(\phi_{k-1}).$$

Qed.

Remark 3 *The bound for $u^tA^{(k)}u/u^tM^{(k)}u$ for version (i) could also have been derived by scaling the matrices $A^{(k)}$ as $A^{(k)} := \mu_k^{-1}A^{(k)}$, which transforms the above analysis into that of Axelsson and Vassilevski [5]. However, this scaling leads to a decidedly inferior lower bound for version (ii):*

$$\phi_{k+1} = \mu_k(1 - \phi_k)$$

Moreover, this scaling would only be of theoretical value. In practice, when the value of μ_k may only be estimated, the scaled method is expected to perform less good than the unscaled method in the presence of inaccurate estimates for μ_k.

Note that the above theorem is somewhat independent of the actual choice of the polynomials; the results hold for any polynomial such that its maximum in the interval I_k is taken at the left boundary.

Also it is worthwile to remark that there is an essential difference in behaviour between polynomials of odd and even degree in that polynomials of even degree are very sensitive to estimates of the upper bound of I_k. While underestimating this bound will for polynomials of odd degree merely result in eigenvalues of $M^{-1}A$ greater than 1, for even degree it may lead to divergent methods.

To analyse the spectral condition number λ_k of ${M^{(k)}}^{-1}A^{(k)}$ for version (i) we note that for $\nu = 1$

$$\lambda_{k+1} = 1/(1 - P_{\nu_k}(\phi_k)) = 1/(\phi_k\mu_k) = \mu_k^{-1}\lambda_k,$$

so the condition number on the finest mesh

$$\lambda_p = \prod_{k=0}^{p-1} \mu_k^{-1}.$$

As $\mu_k^{-1} \geq 1 + \tau$ for some positive τ, the condition number grows geometrically with the number of levels. An upper bound for the number of iterations to achieve a certain relative error ϵ in the norm $\{r^t{M^{(p)}}^{-1}r\}^{1/2}$, where r is the residual, is found to be (see [3])

$$\#iterations = \frac{1}{2}\sqrt{\lambda_p}\log 2/\epsilon.$$

For the standard nine-point stencil we have

$$\lambda_p \to \frac{1}{3}c_1(\sigma^{-1})^p \qquad (c_1 = (5 + 2\sqrt{5})/15, \sigma^{-1} \approx 1.618)$$

so

$$\#iterations = O((1/\sigma)^{p/2}) \log 2/\epsilon = O(2^{(\log_2 1/\sigma)p/2}) \log 2/\epsilon.$$

With $h^{-1} = 2^{p/2}$ we find

$$\#iterations = O(h^{-\log_2 1/\sigma}) \log 2/\epsilon, \qquad \text{for } h \to 0,$$

which incidently shows that the number of iterations grows asymptotically slower than for the incomplete factorization method, but faster than for the modified incomplete factorization method (see [3], for instance).

For version (ii) and for the relative condition of $S^{(k)}$ and $M^{(k)}$ we get a similar analysis, but involving ratios of μ_k's.

Consider now the case $\nu \geq 2$. Actually, as we have seen in the previous section, only $\nu = 2$ is of interest because of the high computational complexity for $\nu \geq 3$. For $\nu = 2$, (12) shows that for version (i)

$$P_{\nu_k}^{(k)}(t) = \left(\frac{\mu_k^{-1} + \phi_k - 2t}{\mu_k^{-1} + \phi_k}\right)^2, \qquad P_{\nu_k}^{(k)}(\phi_k) = \left(\frac{\mu_k^{-1} - \phi_k}{\mu_k^{-1} + \phi_k}\right)^2$$

so

$$\phi_{k+1} = 4\phi_k \mu_k^{-1}/(\mu_k^{-1} + \phi_k)^2, \qquad k = 0, \ldots, p-1. \tag{13}$$

It is readily seen that if $\mu_k \equiv \hat{\mu}$, the recursion converges to a positive fixed-point

$$\hat{\phi} = 2\hat{\mu}^{-1/2} - \hat{\mu}^{-1}$$

if $\hat{\mu} > 1/4$. Even though the sequence μ_k is in fact not constant, the quantities ϕ_k are bounded below by a number independent of p. The condition numbers are

$$\lambda_k = \phi_k^{-1},$$

and

$$\lambda_{k+1} = (1 + \mu_k \lambda_k^{-1})^2/(4\mu_k \lambda_k^{-1}), \qquad k = 0, \ldots, p-1.$$

We have seen earlier that $b_0 = \sigma a_0$ implies $\mu_k = \sigma$, in which case

$$\lambda_k \to \hat{\lambda} = \hat{\mu}/(2\sqrt{\hat{\mu}} - 1) \approx 1.09,$$

i.e., a very small condition number. The condition numbers for the standard nine-point difference stencils converge to the same limit, and should be close to this.

Consider next preconditioners where on alternating levels polynomials of first and higher degree are used, that is

$$\nu_k = 1, \text{ and } \nu_{k+1} = 2 \text{ or } \nu_{k+1} = 3.$$

For $\nu_{k+1} = 2$ we find

$$\begin{aligned} \lambda_{k+2} &= \mu_{k+1}^{-1} \lambda_{k+1} \\ &= \mu_{k+1}^{-1}(1 + \mu_k \lambda_k^{-1})^2/(4\mu_k \lambda_{k+1}^{-1}). \end{aligned} \tag{14}$$

Similar to the above derivation for the case where $\nu = 2$ was used throughout, we find that 14 has a fixed-point

$$\hat{\lambda} = \hat{\mu}/(2\hat{\mu} - 1)$$

if $\mu_k \equiv \hat{\mu}$ and $\hat{\mu} > 1/2$. For $\hat{\mu} = \sigma$ we find

$$\hat{\lambda} = \hat{\sigma}/(2\sigma - 1) = (3 + \sqrt{5})/2 \approx 2.62.$$

For $\nu_{k+1} = 3$ we have

$$\begin{aligned} 1 - P_3(\phi_k) &= \left[T_3\left(\tfrac{\beta_k+\alpha_k}{\beta_k-\alpha_k}\right) - 1\right] \Big/ \left[T_3\left(\tfrac{\beta_k+\alpha_k}{\beta_k-\alpha_k}\right) + 1\right] \\ &= \left[1 + \tfrac{1}{8}\tfrac{(\beta_k-\alpha_k)^2}{(\beta_k+\alpha_k)\beta_k}\right] \Big/ \left[1 + \tfrac{1}{8}\tfrac{(\beta_k-\alpha_k)^2}{(\beta_k+\alpha_k)\alpha_k}\right] \end{aligned}$$

so with $\beta_{k+1} = \mu_{k+1}^{-1}$, $\alpha_{k+1} = \phi_{k+1}$,

$$\begin{aligned} \lambda_{k+2} &= \mu_{k+1}^{-1}\lambda_{k+1} \\ &= \mu_{k+1}^{-1}\left[1 + \frac{1}{8}\frac{(\lambda_k - \mu_k)^2}{(\lambda_k + \mu_k)\mu_k}\right] \Big/ \left[1 + \frac{1}{8}\frac{(\lambda_k - \mu_k)^2}{(\lambda_k + \mu_k)\lambda_k}\right] \end{aligned} \quad (15)$$

If $\mu_k \equiv \hat{\mu}$ this recurrence has a fixed-point $\hat{\lambda}$ satisfying

$$\hat{\lambda}\left[1 + \frac{1}{8}\frac{(\hat{\lambda} - \hat{\mu})^2}{(\hat{\lambda} + \hat{\mu})\hat{\lambda}}\right] = \frac{1}{\hat{\mu}}\left[1 + \frac{1}{8}\frac{(\hat{\lambda} - \hat{\mu})^2}{(\hat{\lambda} + \hat{\mu})\hat{\mu}}\right]$$

and we find if $\hat{\mu} > 1/3$

$$\hat{\lambda} = \hat{\mu}\frac{3 - \hat{\mu}}{3\hat{\mu} - 1}$$

which is

$$\hat{\lambda} = \frac{3}{2} + \frac{\sqrt{5}}{10} \approx 1.72$$

if $\hat{\mu} = \sigma$.

7 Numerical tests

We have tested the method described above on the Poisson problem

$$-\Delta u = 0 \qquad \text{in } \Omega = (0, 1)^2$$

on a unit square with Dirichlet boundary conditions

$$u = 1 \qquad \text{on } \Gamma = \partial\Omega$$

to test the optimality of the method.

The iterative method used is the conjugate gradient method with a multi-level preconditioner $M = M^{(1)}$ with various degrees of polynomials, subject to a stopping criterion

$$\sqrt{g^t M^{-1} g} < 10^{-10}$$

where g is the residual. The coefficient matrix was derived from the standard nine-point difference box.

$1/h =$	10		20		40		80	
$\nu = 1$	10	.31	14	.38	16	.44	19	.50
2	6	.11	6	.09	6	.09	6	.08
3	6	.07	6	.08	6	.08	5	.075
1&2	7	.15	10	.25	9	.23	9	.20
1&3	7	.15	7	.15	7	.14	7	.13

Table 7.1 Numbers of iterations and average reduction rates for the Δ_5/Δ_9 difference stencil on the Poisson problem

The polynomial degrees used were 1, 2, and 3, and first degree alternating with second and third; in the last two cases, the polynomials of first degree were taken on the skew grids.

In table 7.1 we have listed the numbers of iterations (k) and the average reduction rate $(g^t M^{-1} g)^{1/2k}$ We see that the method converges in an optimal number of iterations on the Poisson problem, for polynomial degrees 2 or more. For polynomial degree 1 we see an increase in the number of iterations, although somewhat slower than the discussion in the previous section led us to expect. For a closer discussion of this case see [13].

The use of alternating polynomial degrees degrades performance hardly or not at all, so optimal order preconditioners with a reasonably low number of operations are feasible using the multilevel scheme.

As the computational labour per iteration step for the combination of first and third degree polynomials is more than double that for the combination with first and second degree, the latter combination is to be preferred. The other choices, where polynomials of a single degree are used throughout, have either not an optimal rate of convergence, or not an optimal order of operations per iteration.

References

[1] O. Axelsson, On multigrid methods of the two-level type, in: Multigrid methods, Proceedings, Köln-Porz, 1981, W. Hackbusch and U. Trottenberg, eds., LNM 960, 1982, 352–367.

[2] O. Axelsson, A multilevel solution method for none-point difference approximations, chapter 13 in Parallel Supercomputing: Methods, Algorithms and Applications, Graham F. Carey (ed.), John Wiley, 1989, 191–205.

[3] O. Axelsson, V.A. Barker, Finite element solution of boundary value problems. Theory and computation., Academic Press, Orlando, Fl., 1984.

[4] O. Axelsson, I. Gustafsson, On the use of preconditioned conjugate gradient methods for red-black ordered five-point difference schemes, J. Comp. Physics, 35(1980), 284–299.

[5] O. Axelsson, P. Vassilevski, Algebraic multilevel preconditioning methods I, Numer. Math., 56(1989), 157–177.

[6] O. Axelsson, P. Vassilevski, Algebraic multilevel preconditioning methods II, report 1988-15, Inst. for Sci. Comput., the University of Wyoming, Laramie.

[7] D. Braess, The contraction number of a multigrid method for solving the Poisson equation, Numer. Math., 37(1981), 387–404.

[8] R.E. Ewing, R.D. Lazarov, and P.S. Vassilevski, Local refinement techniques for elliptic problems on cell-centered grids, Report #1988-16, Institute for scientific computation, University of Wyoming, Laramie.

[9] V. Eijkhout and P. Vassilevski, The role of the strengthened Cauchy-Buniakowsky-Schwarz inequality in multi-level methods, submitted to SIAM Review.

[10] T. Meis, Schnelle Lösung von Randwertaufgaben, Z. Angew. Math. Mech, 62(1982), 263–270.

[11] W.F. Mitchell, Unified multilevel adaptive finite element methods for elliptic problems, report UIUCDCS-R-88-1436, Department of Computer Science, the University of Illinois at Urbana-Champaign, Urbana, Illinois, 1988.

[12] M. Ries, U. Trottenberg, G. Winter, A note on MGR methods, Lin. Alg. Appl., 49(1983), 1–26.

[13] P.S. Vassilevski, Nearly optimal iterative methods for solving finite element elliptic equations based on the multilevel splitting of the matrix, Report #1989-01, Institute for scientific computation, University of Wyoming, Laramie.

ITERATION METHOD AS DISCRETIZATION PROCEDURES

O.Axelsson[1] and W. Layton[2]

Abstract:

Much work has been done upon the fast solution of the linear systems arising from upwind difference and central difference approximations to convection dominated convection diffusion equations and yet neither scheme produces acceptable approximate solutions to the original boundary value problem. Still, if these two discretizations are combined via a defect correction approach one obtains an algorithm with many of the best features of both and for which the insights on the fast solution of the associated linear system are very useful. In this report, we describe the algorithm and summarize some of the authors' recent work which gives local error estimates in 2-D on a finite element implementation of this promising scheme.

Key Works Convection-diffusion equation, defect correction method, interior estimates.

1. Introduction. When one solves a boundary value problem for a partial differential equation there are frequently (at least) two stages of discretization: the B.V.P. is replaced by a linear system which is approximated by some iterative procedure, e.g., preconditioned conjugate gradient methods. The error in the computed solution is, of course, bounded above by the sums of the (worst case) errors in each step. However, the computed intermediate approximations can be in fact better than predicted by either step alone! This means, these intermediate steps are considered as defect or deferred correction methods (in which one *stops* at some intermediate step) rather than as iterative methods (in which one continues until some convergence criteria is satisfied). The success of Pereyra's B.V.P. code (see [14] for example) attests to the success of this approach.

Computational experience with this attractive method has outpaced its theoretical justification, especially for singularly perturbed problems possessing boundary and interior layers, and for discretizations involving nonuniform meshes. See, for example, Hemker [7], Ervin and Layton [5], Hemker and Spekreijse [17], and Spekreijse [18], for experiments, the latter two treating (nonlinear) fluid flow problems. Also both for linear and nonlinear problems when approximating boundary or interior layers, we

---

[1] Department of Mathematics, University of Nijmegen, Toernooiveld 5, 6525 ED Nijmegen, The Netherlands.

[2] Department of Mathematics and Statistics, University of Pittsburgh, Pittsburgh, PA 15260, U.S.A.

may need a locally refined finite element mesh at shocks. In this paper we address one aspect of the theoretical foundation, namely: the error in the method for singularly perturbed problems with nonuniform meshes.

To this end, we consider the approximate solution of convection dominated, convection -diffusion equations via defect correction methods. Consider the 2-D equilibrium problem

$$\begin{cases} L_\epsilon u := -\epsilon\Delta u + \underline{v}(\underline{x})\cdot\underline{\nabla} u + q(\underline{x})u = f(\underline{x}), \text{ in } \Omega, \\ u = 0 \text{ on } \Gamma = \partial\Omega, 0 < \epsilon \ll 1 \ |\underline{v}| = 0(1) \text{ with respect } \epsilon \text{ in } \Omega. \end{cases} \tag{1.1}$$

Here $\Omega \subset \mathbf{R}^2$ is a convex polygonal domain, $\underline{v}(\underline{x})$ is a smooth vector field, q and f are smooth functions defined on $\bar{\Omega}$ and the positivity condition

$$q(\underline{x}) - \frac{1}{2}div(\underline{v}(\underline{x})) \geq q_0 > 0, x \in \Omega, \tag{1.2}$$

is assumed to hold throughout. We restrict ourselves to the case $\epsilon < h$, where h denotes a maximum triangle diameter of a general nonuniform triangulation of Ω, as $\epsilon \geq h$ is well understood and the usual finite element or finite difference methods (to which the algorithm reduces) work well in this case.

Equation (1.1) represents a prototypical (linear) model of fluid flow and transport problems. Indeed, one fundamental dilemma, in fluid flow calculations, namely, the choice between schemes which give oscillations at high Reynolds number beginning near layers and penetrating into smooth regions and those giving only first order accuracy even in smooth regions, can be studied accurately in the simpler convection diffusion problem (1.1). Naturally, fluid flow problems possess other complexities as well, but (1.1) is an accepted test problem for investigating these two effects. Alternately, (1.1) can be thought of as an energy equation (u represents temperature, Boland and Layton [16]) or a transport process (u represents concentration) where ϵ plays the role of the inverse of the Prandtl number.

The deferred correction method we study is a finite element implementation of ideas of Hemker [7], [8]. It was first studied in the finite element context in Axelsson and Layton [3, 13] and later the finite difference implementation was analyzed in Ervin and Layton [5, 6]. Letting S^h denote a generic conforming finite element space, $S^h \subset H_0^1(\Omega)$ containing, e.g., piecewise polynomials of degree $\leq k$ defined on some (nonuniform) triangulation of Ω, the defect correction method proceeds as follows. For $j = 1, 2, \cdots, J$ (J "small") we compute $U^j \in S^h$ via:

$$\begin{cases} B_{\epsilon_0}(U^1, w) = (f, w), \forall w \in S^h, \\ B_{\epsilon_0}(U^j - U^{j-1}, w) = (f, w) - B_\epsilon(U^{j-1}, w), \forall w \in S^h, \\ \epsilon_0 = 0(h) \geq h, j = 2, \cdots, J. \end{cases} \tag{1.3}$$

Here B_ϵ is the bilinear form associated with L_ϵ:

$$B_\epsilon(u,w) := \int_\Omega \epsilon \nabla u \cdot \nabla w + (\underline{v} \cdot \nabla u)w + quw \; dx,$$

B_{ϵ_0} denotes B_ϵ with ϵ increased to ϵ_0 and $(\cdot,\cdot)$ denotes the ordinary $L^2(\Omega)$ inner product. Typically, we stop the iteration (1.3) after *only* $J = k+1$ *iterates!*

Since there has been much work on the fast solution of the linear systems arising from the (grossly smeared and inaccurate) first order upwind discretization and the (highly oscillatory) central difference discretization, the finite difference implementation of this algorithm might be of interest since it can be based upon both schemes. (We stress, however, that our analysis presented herein is for the *finite element* case (1.3)). This combination produces good results which neither can do individually. Let $\epsilon_0 = 0(h)$ be chosen so that the usual artificial viscosity, central difference approximation $L^h_{\epsilon_0}$ is a positive type scheme, with the associated discretization matrix M_{ϵ_0} and M-matrix. Let $\tilde{L}^h_\epsilon$ be a higher order (possibly unstable) difference approximation to L_ϵ.

We then initially compute U^1, the $0(h)$ accurate artificial viscosity approximation to u and the later updates U^k via:

STEP1: SOLVE $L^h_{\epsilon_0} U^1 = f$ for U^1.

STEP2: COMPUTE RESIDUAL USING $\tilde{L}^h_\epsilon$: $R^k = f - \tilde{L}^h_\epsilon U^k$

STEP3: UPDATE BY SOLVING: $L^h_{\epsilon_0}(U^{k+1} - U^k) = R^k$

STOP: when $k + 1 = \tilde{k} :=$ the formal order of accuracy of $\tilde{L}^h_\epsilon$.

Clearly, this can be interpreted as: precondition a higher order accurate discretization matrix by the one arising from first order upwind or artificial viscosity. Compute the first few iterations - then stop! The result is better than when iterated to convergence. The basic method, analyzed in [5], [6] for the 1-D problem uses second order central differences for $\tilde{L}^h_\epsilon$. However, the analysis is adaptable to better implementations using fourth order accurate $\tilde{L}^h_\epsilon$ and still better ones basing the discretizations $\tilde{L}^h_\epsilon$ on third and fifth order upwinding methods. It is even possible to adapt, point by point, the accuracy for the stencil used to compute the residuals to reflect some estimate for the smoothness of u, obtained from $U^1, \cdots, U^{j-1}$.

In the nonlinear fluid flow context varying $\epsilon (= Re^{-1})$ leads to varying physical behavior, separation, etc. In the linear problem (1.1) this is not the case. Thus, in the nonlinear problem the choice of ϵ_0 must be refined. ϵ_0 may be picked triangle by triangle, point by point, etc. Alternately, ENO or TVD ideas can be used in the defect correction context, see Hemker, Spijkerse [17].

Our goal here is to describe a very recent result concerning the error in the process (1.3) upon subdomains sufficiently far from Γ. It is observed in both finite difference and finite element implementations of (1.3) that, when unmodified, the error in the method in such subdomains behaves like $0(h^k + (\epsilon_0 - \epsilon)^j)$, where k is the formal order of accuracy of the more accurate discretization. Note that as $\epsilon_0 = 0(h)$ this estimate can be rewritten as error $= 0(h^{min(k,j)})$. Further, it is observed that as j increases we must move to successive subdomains farther from the boundary layers (see [4]) along Γ for the very high order interior estimates above to hold. (Although, modifications exist which seem not to require this.) We shall give a precise statement of the theorem in section 2. Under numerous assumptions upon Ω, $\underline{v}$ etc., we show [see Theorem 2.3, Section 2] that for a subdomain Ω_j', $0(\sqrt{h})$ and $0(h)$ respectively from the characteristic and outflow portions of Γ, the error in H^1 is of the observed form $0(h^k + (\epsilon_0 - \epsilon)^j)$ uniformly in ϵ; up to a term of infinite order accuracy in h which is nonuniform in ϵ, for every $s > 0$:

$$||u - U^1||_{1,\Omega_j'} \leq C_1(s,f)(h^k + (\epsilon_0 - \epsilon)^j) + C_2(\epsilon,j)C_3(s)h^s.$$

This estimate is for the finite element implementation, so no mesh uniformity, etc., assumptions are required, only *local* quasi uniformity in the form of the usual inverse estimates. We also give an $L^2(\Omega_j')$ error estimate which is suboptimal by $0(h^{1/2})$ – typical for these problems.

We treat herein only the most favorable cases for results of the form given above. There are many cases intermediate between Theorem 2.3 and the global estimates given in Axelsson and Layton [3]. For brevity we shall not address ourselves to these cases. Also there are many special cases [notably the case of Courant elements] when the assumptions of Theorem 2.3 can be relaxed or the theorem improved somewhat. Recently, we believe to have improved the proof so as to eliminate the nonuniform $C_2(\epsilon,j)C_3(s)h^s$ term in the estimates.

Note that although on the surface the error estimate above holds in smaller and smaller (as j increases) regions away from layers, these regions extend to the edges of the layers as they are $0(h)$ (a few mesh widths) from outflow type layers and $0(\sqrt{h})$ from characteristic layers. Further, j is really not an iteration number as $j \leq k =$ formal order of the more accurate operator, (for example, the case $k = 2$ is reported in many computational experiments) so the subdomains shrinking slightly with j is not a major effect. There are also computationally attractive modifications of (1.3) which seem in experiments to both sharpen layers and yield the higher accuracy up to their edge.

The notation we use is all standard. $||\cdot||_{k,D}, D \subset \mathbf{R}^2$, denotes the $H^k(D)$ norm, (if $D = \Omega$ then we shall omit D) and $|\cdot|_{k,D}$ the corresponding seminorm. Given a weight function $\psi(x)$, $||w||_{\psi,0}$ denotes the usual weighted $L^2(D)$ norm defined as: $(\int_D w^2\psi\ dx)^{1/2}$. Ω is assumed to be a convex polygon in $\mathbf{R}^2$. We define a finite element space

$S^h = S^h(\Omega) \subset H_0^1(\Omega)$ by first covering Ω with an edge to edge triangulation $\{\tau_\ell\}$ which is assumed to be quasiuniform. We shall assume $S^h|_{\tau_\ell}$ contains polynomials of degree $\leq k$ so that the usual interpolation result holds: for all $u \in H^{k+1}(\Omega) \cap H_0^1(\Omega)$

$$||u - I_h(u)|| + h||u - I_h(u)||_1 \leq Ch^{k+1}|u|_{k+1,\Omega},$$

where $I_h(u) \in S^h$ is the interpolant of u. If D is a subdomain whose boundary consists of edges of the triangulation $\{\tau_\ell\}$ we denote by $S^h(D)$ the restriction of S^h to D. We shall use the quasiuniformity assumption upon $\{\tau_\ell\}$ in the form of the usual inverse estimates for S^h.

For a subdomain Ω, we let Γ_j^+, Γ_j^-, Γ_j° denote, respectively, the inflow, outflow and characteristic portions of $\partial\Omega_j$: $\Gamma_j^+ = \{x \in \partial\Omega_j | \underline{v} \cdot \underline{n}_j > 0\}$, $\Gamma_j^- = \{x \in \partial\Omega_j = \Gamma_j | \underline{v} \cdot \underline{n}_j < 0\}$, $\Gamma_j^\circ = \{x \in \partial\Omega_j = \Gamma_j | \underline{v} \cdot \underline{n}_j = 0\}$. Here $\underline{n}_j$ denotes the outward unit normal to Ω_j.

In Section 2 we review some basic properties of "α-compensating" cutoff functions, introduced by Johnson and Navert [9], Navert [12]. Exponential cutoff functions were earlier used in Galerkin procedures in a different context by Axelsson [2]. We then quote two theorems of Navert [12] which we use extensively in the sequel. With the appropriate formalism introduced, we give the precise statement of the main result of our report in Theorem 2.3. We end the section by recording some local regularity properties of the continuous iteration. The proof of the main theorem naturally breaks into two major steps in which we estimate the continuous iteration error (in section 3) and the discretization error (in section 4).

For other interior estimates for *streamline* diffusion finite element approximations to (1.1) see [9], [10], [11], [12]. Numerical experiments with the finite difference, defect correction method are given in [5], [6] and the work of Hemker [7], [8].

2. Preliminaries. As we will be dealing with interior estimates, it is useful to briefly collect some of the basic properties of the cutoff functions ψ, introduced in Johnson and Navert [9], Navert [12], which we will use. In this work ([9], [12]) the crucial role of subdomains which exclude upstream cutoff was first pointed out and exploited.

Definition 2.1 [Navert [12; p. 19]]. A smooth function ψ is $q_0 = \alpha$-compensating in $\Omega' \subset \Omega$ if $\psi \geq 1$ in Ω' and there is a constant $\rho \in [0, 1]$, with $\rho > 0$ if $q_0 = 0$, such that $-\frac{\partial}{\partial \underline{v}} \psi \geq \rho\psi$ in Ω. Note that q_0 is defined in (1.2).

Definition 2.2. A subdomain $\Omega'' \subset \Omega$ is said to "exclude upstream cutoff" if all points upstream w.r.t. $\underline{v}$ of a point in Ω'' belong to Ω''. I.e., $(\Gamma'')^- \subset \Gamma$.

In Navert [12; L. 2.1 p. 20] and Johnson and Navert [9], appropriate cutoff functions are constructed for subdomains which exclude upstream cutoff. We recall, for later reference, their properties by quoting the following result:

Lemma 2.1 [Navert [12; Lemma 2.1, pp. 20, 21]]

Let Ω'' be a subdomain of Ω with piecewise smooth boundary Γ'' such that $(\Gamma'')^- \subset \Gamma$. Let c, s, d, γ and M be positive numbers with $d \leq 1/2$, $\gamma \geq 1$ and M an integer. Assume that for all points upstream w.r.t. $\underline{v}$ a point on $(\Gamma'')^0$ lie on $(\Gamma'')^0$ and that $|\underline{v}\cdot\underline{n}''| \geq c$ on $(\Gamma'')^- \cup (\Gamma'')^+$, where $\underline{n}''$ is the outward unit normal to Ω''. Then, there are constants $C_1 = C_1(s, M, \underline{v}, \Omega'')$, $C_2 = C_2(s, M, \underline{v}, \Omega')$ and $C = C(\underline{v}, \Omega'')$ such that if Ω' is any subdomain of Ω'' at a distance of at least $C_1\gamma d \ell n(\frac{1}{d})$ and $C_2\gamma\sqrt{d}\,\ell n(\frac{1}{d})$ from $(\Gamma'')^+$ and $(\Gamma'')^0$ respectively, there exists a cutoff function ψ which is $q_0 = \alpha$ compensating in Ω' with the following properties:

$$\psi \geq 1 \text{ in } \Omega', \tag{2.1;a}$$

$$\psi = 0 \text{ in } \Omega - \Omega'', \tag{2.1;b}$$

$$\frac{\partial}{\partial \underline{v}}\psi \leq 0, \text{ in } \Omega, \tag{2.1;c}$$

$$\max_{|\underline{y}|\leq\gamma d} \frac{\psi(\underline{x}+\underline{y})}{\psi(\underline{x})} \leq \mathrm{C}, \text{ in } \{\underline{x}\in\Omega|\psi(\underline{x}) \geq \mathrm{C}d^s\}, \tag{2.1;d}$$

$$\max_{|\underline{y}|\leq\gamma d} \frac{-\psi_{\underline{v}}(\underline{x}+\underline{y})}{-\psi_{\underline{v}}(\underline{x})} \leq \mathrm{C}, \text{ in } \{\underline{x}\in\Omega| -\psi_{\underline{v}}(\underline{x}) \geq \mathrm{C}d^s\}, \tag{2.1;e}$$

$$|D^m_{\underline{v}}\psi| \leq C\{\gamma^{-1}d^{-m}\psi + d^s\}, \text{ in } \Omega \text{ if } 0 < m \leq M, \tag{2.1;f}$$

For any $\underline{v}^\perp$ with $\underline{v}^\perp \cdot \underline{v} = 0$, $|\underline{v}^\perp| = 0(1)$,

$$|D^\gamma_{\underline{v}^\perp}\psi| \leq C\{\gamma^{-1}d^{-1/2|\lambda|}\psi - d^{1-|\lambda|}\psi_{\underline{v}} + d^s\}, \text{ in } \Omega, \tag{2.1;g}$$

if $0 < |\lambda| \leq M$,

$$|D_{\underline{v}}^{m+1} D_{\underline{v}^{\perp}} \psi| \leq C\{-\gamma^{-1} d^{-m-|\lambda|} \psi_{\underline{v}} + d^{s}\}, \text{ in } \Omega, \tag{2.1; h}$$

if $0 < m + |\lambda| < M$.
Here D_ξ denotes differentiation in the ξ direction, $D_\xi w = \xi \cdot \nabla w$ etc., $m \geq 0$ is an integer and $\lambda = (\lambda_1, \cdots, \lambda_n)$ is a multi-index of length $|\lambda| = \Sigma \lambda_j$.

In the sequel we shall always make the assumption that

$$\Omega_j', \Omega_j'' \text{ satisfy the hypotheses of Lemma 2.1 with } \mathrm{d} = \epsilon_0 = 0(\mathrm{h}). \tag{2.2}$$

Definition 2.3 [Navert [12; Def. 2.2 p. 25]] A cutoff function ψ satisfying the conditions of Lemma 2.1 is said to be $q_0 = \alpha$ compensating and smooth of order (d, γ, s) on (Ω', Ω'').

We next quote two basic results of Navert [12] which are used extensively in the sequel. The first is a local regularity result for the continuous boundary value problem. The second can be interpreted as a local error estimate upon the artificial viscosity projection operator.

Theorem 2.1 [Navert [12; Theorem 2.3 p. 25]] Assume that either $\underline{v}$ has no closed arcs in $\bar{\Omega}$ or that $q_0 > \tilde{C}(\underline{v})$ where $\tilde{C}$ depends upon the first derivatives of $\underline{v}$. Let Ω', Ω'' satisfy Lemma 2.1, let c, s, m be positive numbers with m an integer and let $(\Gamma'')^- \subset \Gamma^-$ and Γ^- is part of the faces of Ω (a polygon). Suppose Ω is a convex polygon in $\mathbf{R}^2$, $\overline{(\Gamma'')}^- \cap \bar{\Gamma}^+$ is empty. Then, there are constants $C_1 = C_1(s, m, \underline{v}, \Omega'')$, $C_2 = C_2(s, m, \underline{v}, \Omega'')$ and $C = C(\underline{v}, q, \Omega'', \Omega)$ such that if the distance from Ω' to $(\Gamma'')^+$ and $(\Gamma'')^0$ is at least $C_1 \epsilon_0 \ell n(\frac{1}{\epsilon_0})$ and $C_2 \sqrt{\epsilon_0} \ell n(\frac{1}{\epsilon_0})$ respectively, the solution of:

$$L_{\epsilon_0} u_{\epsilon_0} = f \text{ in } \Omega, \mathrm{w} = \mathrm{g} \text{ on } \Gamma, \tag{2.3}$$

with $f \in H^m(\Omega'')$ and $g = 0$, satisfies

$$||u||_{m,\Omega'} + ||u_{\underline{v}}||_{m,\Omega'} \leq C(||f||_{m,\Omega''} + \epsilon_0^s ||f||_\Omega). \tag{2.4}$$

Letting u_{ϵ_0} denote the solution to (2.3), with $\epsilon_0 \geq h$ define $U_{\epsilon_0} \in S^h$ by

$$\begin{cases} B_{\epsilon_0}(u_{\epsilon_0} - U_{\epsilon_0}, w) = 0, \forall w \in S^h \cap H_0^1(\Omega), \\ U_{\epsilon_0} = I_h(g), \text{ on } \Gamma. \end{cases} \tag{2.5}$$

Thus, U_{ϵ_0} is the usual Galerkin projecton of u_{ϵ_0} into S^h. (2.5) can also be thought of as the Galerkin artificial viscosity projection of u_{ϵ_0} into S^h, as $\epsilon_0 \geq h$. We shall frequently use the following result of Navert:

Theorem 2.2 [Navert [12; Corollary 4.3, p. 52]] Let $\epsilon_0 \geq h$, $\Omega' \subset \Omega'' \subset \Omega$ be as in Lemma 2.1, and $u \in H^{m+1}(\Omega)$, $g \in H^{m+1}(\Gamma)$, $0 \leq m \leq k$, integers. Let $s > 0$ be given. Then, there are constants $C_1 = C_1(s, \underline{v}, \Omega'')$, $C_2 = C_2(s, \underline{v}, \Omega'')$ and $C = C(\underline{v}, q, \Omega'', \Omega, T)$ such that if the distance from Ω' to $(\Gamma'')^0$ and $(\Gamma'')^+$ is at least $C_1\sqrt{\epsilon_0}\ell n(\frac{1}{\epsilon_0})$ and $C_2\epsilon_0\ell n(\frac{1}{\epsilon_0})$, resp., then

$$||u_{\epsilon_0} - U_{\epsilon_0}||_{1,\Omega'} \leq C\{h^k|u_{\epsilon_0}|_{k+1,\Omega''} + \epsilon_0^s h^m |u_{\epsilon_0}|_{m+1,\Omega}\}. \tag{2.6; a}$$

Moreover, if Ω' is convex, Γ'' consists of gridlines and $u|_{\Gamma''} \in H^{k+1}(\Gamma'')$ then,

$$\begin{aligned}||u_{\epsilon_0} - U_{\epsilon_0}||_{\Omega'} \leq C\{&\frac{h^{k+1}}{\epsilon_0}(|u_{\epsilon_0}|_{k+1,\Omega''} + |u_{\epsilon_0}|_{k+1,\Gamma''}) \\ &+ \epsilon_0^s h^{m+1}(|u_{\epsilon_0}|_{m+1,\Omega} + |g|_{m+1,\Gamma})\}.\end{aligned} \tag{2.6; b}$$

We shall now state the main theorem. Let

$$\Omega = \Omega_0'' = \Omega_0' \supset \Omega_1'' \supset \Omega_1' \supset \cdots \supset \Omega_j'' \supset \Omega_j' \tag{2.7; a}$$

be a sequence of subdomains of Ω such that

each Ω_j'', Ω_j' satisfy the hypotheses of Lemma 2.1 with $d = \epsilon_0 \geq h$ and does not admit upstream cutoff, $j = 1, 2, \cdots, J$. (2.7; b)

Further suppose Γ_j'', Γ_j' satisfy condition (3.3a, b, c) of section 3. (2.7; c)

Suppose also:

$\underline{v}$ has no closed arcs in $\bar{\Omega}$ or $q_0 > \tilde{C}(\underline{v})$ where $\tilde{C}$ depends upon the first derivatives of $\underline{v}$, (2.7; d)

$(\overline{\Gamma_j''})^- \cap \overline{\Gamma^+}$ is empty, (2.7; e)

Ω_j'' satisfies conditions (4.1a, b), (given in section 4), (2.7; f)

Γ''_j consists of grid lines [necessary for L^2estimate]. (2.7; g)

Remark: The constraint that Γ''_j consist of grid lines in (2.7;g) is not necessary for the H^1 estimate in Theorem 2.3 below, only for the L^2 estimate. This requirement can also be omitted if the finite element space consists of C° piecewise linears on triangles (Courant elements).

Theorem 2.3. Suppose that (2.7;a) through (2.7;f) above hold. Then, for all $s > 0, j = 1, \cdots, J$, it follows that for $r = 0$:

$$\epsilon_0^{1/2}||u - U^j||_{1,\Omega'_j} + r||u - U^j||_{\Omega'_j}$$
$$\leq C(\epsilon_0 - \epsilon)^j[\epsilon_0^{1/2}||f||_{2j+1,\Omega'_0} + r||f||_{2j,\Omega'_0}]$$
$$+ Ch^{k+1/2}[\sum_{p=1}^{j-1}\sum_{\ell=0}^{p-1}(\epsilon_0 - \epsilon)^\ell||f||_{k+2+2\ell,\Omega'_{p-\ell-1}}]$$

$$+C\epsilon_0^s[((\epsilon_0 - \epsilon)^{j+1}(\epsilon^{3/2})^{j+1} + 1)||f||_{0,\Omega} + \sum_{p=1}^{j-1}\sum_{\ell=0}^{p-1}(\epsilon_0 - \epsilon)^\ell||f||_{2+2\ell,\Omega''_{p-\ell-1}}]$$

If (2.7;g) holds additionally then the above holds for $r = 1$.

We shall prove Theorem 2.3 in two steps in, resp., Theorem 3.1 section 3 and Theorem 4.1 section 4.

We conclude this section by collecting some useful local regularity properties of a continuous analogue of the sequence U^j. Define the continuous iterate s u^j by: $u^0 = 0$ and

$$L_{\epsilon_0}u^j = f - (\epsilon_0 - \epsilon)\Delta u^{k-1} \text{ in } \Omega, u^j = 0, \text{ on } \partial\Omega. \tag{2.8}$$

Theorem 2.4 Suppose (2.7;a-g) holds. Then, for any positive integer m and positive number s, there are constants $C_1 = C_1(s, m, \underline{v}, \Omega''_j)$ the solutions u^j of (2.8) satisfy for $j = 1, \cdots, J < \infty$,

$$||u^j||_{m,\Omega'_j} + ||u^j_{\underline{v}}||_{m,\Omega'_j} \leq C\{\sum_{\ell=0}^{j-1}(\epsilon_0 - \epsilon)^\ell||f||_{m+2\ell,\Omega''_{j-1}} + \epsilon_0^s||f||_\Omega$$
$$+ \epsilon_0^{s+1}\sum_{\ell=1}^{j-1}||\Delta u^\ell||_\Omega\}.$$

Proof: Note that by Navert [12; Theorem 2.3, p. 26]

$$||u^1||_{m,\Omega_1'} + ||u^1_{\underline{v}}||_{m,\Omega_1'} \leq C(||f||_{m,\Omega_1''} + \epsilon_0^s||f||_\Omega).$$

Similarly,

$$|u^j||_{m,\Omega_j'} + ||u^j_{\underline{v}}||_{m,\Omega_j''} \leq \epsilon_0^s||f||_\Omega + (\epsilon_0 - \epsilon)(||u^{j-1}||_{m+2,\Omega_j''} + \epsilon_0^s||\Delta u^{j-1}||_\Omega)\}.$$

Thus, by backwards recursion:

$$||u^j||_{m,\Omega_j'} + ||u^j_{\underline{v}}||_{m,\Omega_j'} \leq C(\sum_{\ell=0}^{j-1}(\epsilon_0 - \epsilon)^\ell||f||_{m+2\ell,\Omega_{k-1}''}$$
$$+ \epsilon_0^s||f||_\Omega + (\epsilon_0 - \epsilon)\epsilon_0^s\sum_{\ell=1}^{j-1}||\Delta u^\ell|| + \Omega)$$

Analogously, using instead Axelsson and Layton [3; Lemma 2.1] we derive global estimates for u^j.

Proposition 2.1. suppose $q_0 > 0$ and Ω is a regular domain. Then for $1 \leq j$

$$||\Delta u^j|| \leq C\epsilon_0^{-1-j/2}||f||$$

Proof: From [3; Lemma 2.1 equation (2.2;b)]

$$||\Delta u^1|| \leq C\epsilon_0^{-3/2}||f||.$$

Similarly

$$||\Delta u^j|| \leq C\epsilon_0^{-3/2}[||f|| + (\epsilon_0 - \epsilon)||\Delta u^{j-1}||] \leq \cdots \leq C\epsilon_0^{-3/2}\sum_{\ell=0}^{j-1}(\epsilon_0^{-1/2})^\ell||f||,$$

and the result follows.

Combining the previous two results gives.

Corollary 2.1 Under the assumptions (2.7;a) - (2.7;f) of Theorem 2.3, for $j = 1, \cdots, J < \infty$,

$$||u^j||_{m,\Omega_j'} + ||u^j_{\underline{v}}||_{m,\Omega_j'} \leq C\{\sum_{\ell=0}^{j-1}(\epsilon_0 - \epsilon)^\ell||f||_{m+2/ell,\Omega_{j-1}''}$$
$$+ \epsilon_0^s[1 + (\epsilon_0^{-1/2}\epsilon_0^{\frac{-(i-1)}{2}})]||f||_\Omega\}$$

and for $j = 1, \cdots, J < \infty$

$$\|u^j\|_{m,\Omega_j''} + \|u_{\underline{v}}^j\|_{m,\Omega_j''} \leq C\{\sum_{\ell=0}^{j-1}(\epsilon_0 - \epsilon)^\ell \|f\|_{m+2\ell,\Omega_{j-\ell-1}}$$
$$+ \epsilon_0^s[1 + (\epsilon_0^{-1/2}\epsilon_0^{-(\frac{j-1}{2})})]\|f\|_\Omega\}.$$

3. The Iteration Error. Letting u^j denote the continuous defect correction iteration, $\delta^j := u - u^j$ is easily seen to satisfy the following:

$$\begin{cases} L_{\epsilon_0}\delta^j = (\epsilon_0 - \epsilon)(-\Delta\, \delta^{j-1}), \text{ in } \Omega, \\ \delta^j = 0 \text{ on } \partial\Omega, \delta^0 = \mathrm{u}, \mathrm{j} = 1, \cdots, \mathrm{J}. \end{cases} \tag{3.1}$$

We shall derive interior estimates for δ^j upon certain subsets Ω_j', Ω_j'' of Ω. Let

$$\Omega = \Omega_0'' \supset \Omega_0' \supset \Omega_1'' \supset \Omega_1' \supset \Omega_2'' \supset \Omega_2' \supset \cdots \supset \Omega_J'' \supset \Omega_J', \tag{3.2}$$

where each Ω_j', Ω_j'' does not admit upstream cutoff. Specifically, assume that

the boundaries of Γ_j', Γ_j'' are piecewise smooth with (3.3; a)
$(\Gamma_j'')^- \subset (\Gamma_{j-1}')^- \subset \Gamma^-, j = 0, \cdots J$
all points upstream of a point on $(\Gamma_j'')^0$belong $(\Gamma_j'')^0$. (3.3; b)
$|\underline{v} \cdot \underline{n}_j''| \geq c$ on $(\Gamma_j'')^- \cup (\Gamma_j'')^+$
where $\underline{n}_j$ is the outward unit normal to Ω_j'. (3.3; c)

The basic estimate for the iteration error is then:

Theorem 3.1. Suppose (3.2), (3.3;a,b,c) hold. Suppose also that $\underline{v}$ has no closed arcs in $\bar{\Omega}$ or that $q_0 > \tilde{C}(\underline{v})$ where $\tilde{C}(\underline{v})$ depends upon the first derivatives of $\underline{v}$. Assume $(\Gamma_j^{\overline{n}})^- \cap \bar{\Gamma}^+$ is empty and $s > 0$. Then, there are constants $C_1 = C_1(s, \underline{v}, \Omega_j'')$ $C_2 = C_2(x, \underline{v}, \Omega_j'')$ and $C = C(\underline{v}, q, \Omega_j'', \Omega, J)$ such that if

$$\mathrm{dist}(\Omega_j', (\Gamma_j'')^+) \geq C_1\epsilon_0 \ell n\left(\frac{1}{\epsilon_0}\right), \mathrm{dist}(\Omega_j', (\Gamma_j'')^0) \geq C_2\sqrt{\epsilon_0}\ell n\left(\frac{1}{\epsilon_0}\right),$$

it follows that for $m = 0,1,2$

$$||\delta^j||_{m,\Omega_j'} + ||\frac{\partial}{\partial \underline{v}}\,\delta^j||_{m,\Omega_j'} \leq C\{(\epsilon_0-\epsilon)^j||f||_{2j+m,\Omega_0'} + \epsilon^s||f||_{0,\Omega}\}$$
$$+ C\epsilon_0^s(\epsilon_0-\epsilon)^{j+1}\{(\epsilon^{-3/2})^{j+1}||f||_{0,\Omega}\}, j = 1,\cdots,J.$$

Proof: Applying Navert [12, Theorem 2.3 p. 26], which is given in Section 2, Theorem 2.1, to the iteration (3.1) upon the subsets Ω_j'; Ω_j'' of Ω gives: (for $m = 0,1,2$)

$$||\delta^j||_{m,\Omega_j'} + ||\frac{\partial}{\partial \underline{v}}\,(\delta^j)||_{m,\Omega_j'}$$
$$\leq C[(\epsilon_0-\epsilon)||\delta^{j-1}||_{2+m,\Omega_j''} + \epsilon_0^s(\epsilon_0-\epsilon)||\Delta\delta^{j-1}||_{\Omega}]$$
$$\leq \text{ (iterating backward to } \delta^0 = u) \cdots$$
$$\leq (C(\epsilon_0-\epsilon)^j\{||f||_{2j+m,\Omega_0'} + \epsilon^s||f||_{0,\Omega}\}$$
$$+ C\ \epsilon_0^s(\epsilon_0-\epsilon)(\sum_{\ell=0}^{j}||\Delta\delta_\ell||_{\Omega}).$$

The last term is bounded using the global estimate for the iteration error given in Corollary 4.2 of Axelsson and Layton [3]. Namely,

$$||\Delta\delta^\ell||_{0,\Omega} \leq C\ \epsilon^{-3/2}(\frac{\epsilon_0-\epsilon}{\epsilon^{3/2}})^\ell||f||.$$

Thus, as $0 \leq \ell \leq j \leq J$, $\Sigma||\Delta\delta^\ell|| \leq C\ \epsilon^{-3/2}(\frac{\epsilon_0\epsilon}{\epsilon^{3/2}})^j||f||$ and the result follows.

The above theorem shows that, up to a term which is of infinite order accuracy in $\epsilon_0 = 0(h)$ nonuniformly in ϵ the iteration error $u - u_j$ is uniformly in ϵ of order $(\epsilon_0-\epsilon)^j$ upon subdomains Ω_j sufficiently far from Γ^0 and Γ^+. It is noteworthy that, to achieve this accuracy we must move further inside Ω as we iterate further. This increase in the pollution as we iterate was noticed by Hemker [7]. See also Ervin and Layton [6; section 4], for an example in which this effect appears. In 1-D the estimate in Theorem 3.1 can be sharpened considerably [6].

4. The Discretization Error. In this section, we give bounds upon the discretization error $u^j - U^j$ for certain subdomains Ω_j' of Ω. We let:

$\Omega = \Omega_0'' \supset \Omega_1'' \supset \cdots \supset \Omega_j' \supset \Omega_j''$ be a chain of subdomains of Ω (as in section 3) satisfying (3.2), (3.3). Additionally we shall require:

Ω_j'' is convex and $\Gamma_j'' = \partial\Omega_j''$ consists of gridlines of the mesh covering Ω.

$$(4.1;a)$$

$$\text{dist }(\Omega_j'; (\Gamma_j'')^+) \geq C_1\ \epsilon_0 \ell n\left(\frac{1}{\epsilon_0}\right), \text{ and}$$
$$\text{dist }(\Omega_j'; (\Gamma_j'')^0) \geq C_2\sqrt{\epsilon_0}\ell n\left(\frac{1}{\epsilon_0}\right). \tag{4.1;b}$$

Theorem 4.1. Suppose the hypotheses of Lemma 2.1 are satisfied by Ω_j and (3.2), (3.3) (4.1;a,b) hold. Suppose $f(\underline{x})$ is sufficiently smooth ($f \in H^m(\Omega)$ for m sufficiently large). $\epsilon_0 \geq h$, and suppose that each Ω_j', Ω_j'' satisfies the hypotheses of Theorems 2.1 and 2.2. In particular: $(\Gamma_j'')^- \subset \Gamma^-, \overline{\Gamma_j'')^- \cap \overline{\Gamma^+}}$ is empty, Γ^- is part of the faces of Ω and Γ_j'' consists of grid lines. Let $|||\cdot|||_{\Omega_j'}$ be defined as in the proof below. Then, for any $s > 0$ there is a $C(s)$ such that

$$|||u^j - U^j|||_{\Omega_j'} \leq Ch^{k+1/2}\sum_{p=1}^{j-1}\sum_{\ell=0}^{p-1}(\epsilon_0-\epsilon)^\ell||f||_{k+2+2\ell,\Omega''_{p-\ell-1}} +$$
$$c\ \epsilon_0^s[\sum_{p=1}^{j-1}\sum_{\ell=0}^{p-1}(\epsilon_0-\epsilon)^\ell||f||_{2+2\ell,\Omega_{p-\ell-1}} + ||f||_\Omega] \tag{4.2}$$

In particular, for any $s > 0$ there are constants $\tilde{C}_1 = \tilde{C}_1\ (s, f, C_1, C_2)$ $\tilde{C}_2 = \tilde{C}_2\ (f, C_1, C_2)$ such that

$$||u^j - U^j||_{1,\Omega_j'} \leq \tilde{C}_1(s, f, C_1, C_2)h^k + \tilde{C}_2(f, C_1, C_2)\epsilon_0^s, \tag{4.3;a}$$

and

$$||u^j - U^j||_{\Omega_j'} \leq \tilde{C}_1(s, f, C_1, C_2)h^{k+1/2} + \tilde{C}_2(f, C_1, C_2)\epsilon_0^s. \tag{4.3;b}$$

Proof: For the proof we shall require the following norms on the domain Ω and subdomains $\tilde{\Omega}$.

$$|||w|||^2 := \epsilon_0||\nabla w||^2 + ||w||^2, |||w|||_\psi^2 := \epsilon_0||\nabla w||_\psi^2 + ||w||_\psi^2, ||w||_\psi^2 = \int_\Omega w^2\psi dx,$$

$$|||w|||_\psi^2 := \epsilon_0||\nabla w||^2_{L^2(\tilde{\Omega})} + ||w||^2_{L^2(\tilde{\Omega})}, |||w|||_\psi^2 := \epsilon_0||\nabla w||^2_{\psi,\tilde{\Omega}} + ||w||^2_{\psi,\tilde{\Omega}}.$$

Further, we define the operator $\mu_j : H^1(\Omega) \to S^h(\Omega) \cap \{\phi | \phi = I_h(u)$ on $\Omega - \Omega_j''\}$ by $\mu_j(u) = \tilde{w}$ where $\tilde{w}$ satisfies: for all $w \in S^h(\Omega_j'') \cap H_0^1(\Omega_j'')$

$$\begin{cases} 0 = B_{\epsilon_0,\Omega_j''}(u - \tilde{w}, w) := \int_{\Omega_j''} \epsilon_0 \nabla(u - \tilde{w}) \cdot \nabla w \\ \qquad + \underline{v} \cdot \nabla(u - \tilde{w}) w + q(u - \tilde{w}) w dx \\ \tilde{w} = I_h(u) \text{ (interpolant) on } \Omega - \Omega_j'' \end{cases} \tag{4.4}$$

When $j = 1$ we have: $B_{\epsilon_0}(u^1 - U^1, w) = 0$, for all $w \in S^h(\Omega)$, $\epsilon_0 = 0(h)$.
Thus from, e.g., Navert [12; Corollary 4.3 p. 52; Theorem 4.8 p. 49] we have, for integer m, $0 \le m \le k$,

$$||u^1 - U^1||_{1,\Omega_1'} \le C\{h^k |u^1|_{k+1,\Omega_1''} + \epsilon_0^s h^m |u^s|_{m+1,\Omega}\}, \tag{4.5; a}$$

$$||u^1 - U^1||_{1,\Omega_1'} \le C\{h^k |u^1|_{k+1,\Omega_1''} + \epsilon_0^s ||f||_{0,\Omega}\}, \tag{4.5, b}$$

$$||u^1 - U^1||_{0,\Omega_1'} \le C\{h^{k+1/2}(|u^1|_{k+1,\Omega_1''} + |u^1|_{k+1,\Gamma_1''}) + \epsilon_0^s ||f||_{0,\Omega}\} \tag{4.6; a}$$

$$\begin{aligned} ||u^1 - U^1||_{0,\Omega_1'} \le C\{h^{k+1/2}(|u^1|_{k+1,\Omega_1''} + |u^1|_{k+1,\Gamma_1''}) + \epsilon_0^s h^{m+1} \\ (|u^1|_{m+1,\Omega} + |u^1|_{m+1,\Gamma})\}. \end{aligned} \tag{4.6; b}$$

We now proceed by induction; we shall show that, loosely speaking, $|||U^j - u^j|||_{\Omega_j'}$ is of the same order as $|||U^{j-1} - u^{j-1}|||_{\Omega_{j-1}'}$. Define $\Theta_j = \mu_j(u^j) - U^j$ so that

$$|||u^j - U^j|||_{\Omega_j'} \le |||u^j - \mu_j(u^j)|||_{\Omega_j'} + |||\Theta_j|||_{\Omega_j'}, \tag{4.7}$$

and Θ_j satisfies for all $w \in \overset{\circ}{S}{}^h(\Omega_j'')$

$$B_{\epsilon_0,\Omega_j''}(\Theta_j, w) = (\epsilon_0 - \epsilon)(\nabla(u^{j-1} - U^{j-1}), \nabla w)_{\Omega_j''}. \tag{4.8}$$

Introduce the cutoff function ψ which is $q_0 = \alpha$ compensating of order (ϵ_0, γ, s) in (Ω_j', Ω_j''). Note that $\psi H_j = 0$ on Γ^- so we obtain, following, e.g., Navert [12], Johnson, Navert, Pitkaranta [10],

$$B_{\epsilon_0,\Omega_j''}(\Theta_j, \psi\Theta_j) \ge \epsilon_0 ||\nabla\Theta_j||_\psi^2 + \epsilon_0(\nabla\Theta_j, \Theta_j \nabla\psi) + ||\Theta_j||^2_{q_0\psi - \frac{1}{2}\frac{\partial}{\partial \underline{v}}\psi}.$$

Hence, for γ sufficiently large, we find, as in [10][12],

$$B_{\epsilon_0,\Omega_j''}(\Theta_j, \psi\Theta_j) \ge C_1(|||\Theta_j|||_\psi^2 + ||\Theta_j||^2_{-\psi_{\underline{v}}}) - C_2\epsilon_0^s |||\Theta_j|||^2_{\Omega_j''}$$

Therefore,

$$C(\epsilon_0^s|||\Theta_j|||^2_{\Omega''_j} + B_{\epsilon_0,\Omega''_j}(\Theta_j,\psi\Theta_j)) \geq |||\Theta_j|||^2_\psi + ||\Theta_j||^2_{-\psi_{\underline{v}}}. \qquad (4.9)$$

From the equation (4.8) we get, $B_{\epsilon_0,\Omega''_j}(\Theta_j, I_h(\psi\Theta_j)) - (\epsilon_0-\epsilon)(\nabla(u^{j-1} - U^{j-1}), \nabla(I_h(\psi\Theta_j)))_{\Omega''_j} = 0$, so that

$$\begin{aligned} B_{\epsilon_0,\Omega''_j}(\Theta_j,\psi\Theta_j) &\leq B_{\epsilon_0,\Omega''_j}(\Theta_j,(\psi\Theta_j) - I_h(\psi\Theta_j)) - (\epsilon_0-\epsilon) \\ &\quad (\nabla(u^{j-1} - U^{j-1}), \nabla(I_h(\psi\Theta_j)))_{\Omega''_j} \\ &\leq -(\epsilon_0-\epsilon)(\nabla(u^{j-1} - J^{j-1}), \nabla(I_h(\psi\Theta_j)))_{\Omega''_j} \\ &\quad + C\gamma^{-1/2}\{\epsilon_0^{1/2}||\nabla\Theta_j||_\psi + \frac{h}{\epsilon_0}(||\frac{\partial}{\partial \underline{v}}\Theta_j||_\psi + ||\Theta||_\psi)\} \\ &\quad \cdot (||\Theta||_\psi + ||\Theta||_{-\psi_{\underline{v}}}). \end{aligned} \qquad (4.10)$$

Inserting (4.10) into (4.9) gives

$$\begin{aligned} |||\Theta_j|||^2_\psi + ||\Theta_j||^2_{-\psi_{\underline{v}}} &\leq -(\epsilon_0-\epsilon)(\nabla(u^{j-1} - U^{j-1}), \nabla(I_h(\psi\Theta_j)))_{\Omega''_j} + \\ &\quad + C\epsilon_0^s|||\Theta_j|||^2_{\Omega''_j} + C\gamma^{-1}\{\epsilon_0||\nabla\Theta_j||^2_\psi + \frac{h^2}{\epsilon_0}(||\frac{\partial}{\partial \underline{v}}\Theta_j||^2_{psi} + ||\Theta_j||^2_\psi\}. \end{aligned} \qquad (4.11)$$

As $\epsilon_0 \geq 0(h)$, taking γ sufficiently large to hide the appropriate terms in (4.11) gives

$$\begin{aligned} |||\Theta_j|||^2_{\Omega'_j} \leq C(|||\Theta_j|||^2_\psi + ||\Theta||^2_{\psi_{\underline{v}}}) &\leq C\ \epsilon_0^s|||\Theta_j|||^2_{\Omega''_j} \\ &\quad + C(\epsilon_0-\epsilon)|(\nabla(u^{j-1} - U^{j-1}), \nabla(I_h(\psi\Theta_j)))_{\Omega''_j}|. \end{aligned} \qquad (4.12)$$

We now focus our attention on the last term in the R.H.S. of (4.12). Expanding:

$$(\epsilon_0-\epsilon)(\nabla(u^{j-1} - U^{j-1}), \nabla(I_h(\psi\Theta_j)))_{\Omega''_j} = (\epsilon_0-\epsilon)(T_1+T_2) \qquad (4.13)$$

where $T_1 := (\nabla(u^{j-1} - U^{j-1}),\ \nabla(\psi\Theta_j))_{\Omega''_j}$,

$T_2 := (\nabla(u^{j-1} - U^{j-1}), \nabla\{I_h(\psi\Theta_j) - \psi\Theta_j\})_{\Omega''_j}$. Consider T_1 :

$$\begin{aligned} T_1 &\leq ||\nabla(u^{j-1} - U^{j-1}||_\psi||\nabla\Theta_j||_\psi + (\nabla(u^{j-1} - U^{j-1}), \Theta_j\nabla\psi)_{\Omega''_j} \\ &\leq ||\nabla(u^{j-1} - U^{j-1})||_\psi(||\nabla\Theta_j||_\psi + ||\Theta_j\nabla\psi||_{\psi^{-1},\Omega''_j}). \end{aligned}$$

With $\underline{v}^{\perp}$ the velocity field orthogonal to $\underline{v}$:

$$\|\Theta_j \nabla\psi\|_{\psi^{-1},\Omega_j''} \leq C\{\|\Theta_j \frac{\partial}{\partial \underline{v}}\psi\|_{\psi^{-1},\Omega_j''} + \|\Theta_j \frac{\partial}{\partial \underline{v}^{\perp}}\psi\|^2_{\psi^{-1},\Omega_j''}\}$$
$$\leq C\{(\gamma h)^{-1}\|\Theta_j\|^2_{-\psi_{\underline{v}}} + \gamma^{-1}h^{-1}\|\Theta_j\|^2_{\psi_{\underline{v}}}\} + Ch^s\|\Theta_j\|^2_{\psi}, \tag{4.15; a}$$

where we have used the properties of ψ given in Navert [12] see also section 2, Lemma 2.1.

Inserting (4.15) into (4.14) gives the following bound for T_1 of (4.13);

$$T_1 \leq \|\nabla(u^{j-1} - U^{j-1})\|_{\psi}\{\|\nabla\Theta_j\|_{\psi} + C(\gamma h)^{-1/2}\|\Theta_j\|_{-\psi_{\underline{v}}} + Ch^s\|\Theta_j\|_{\psi}\} \tag{4.15; b}$$

We bound T_2 using similar technique as in (4.10), (4.11). Indeed,

$$T_2 \leq \|\nabla(u^{j-1} - U^{j-1})\|_{\psi}\|\nabla(\psi\Theta - I_h(\psi\Theta))\|_{\psi^{-1},\Omega_j''},$$

By interpolation theory:

$$\|\nabla((\psi\Theta_j) - I_h(\psi\Theta_j))\|_{\psi^{-1},\Omega_j''}$$
$$\leq Ch^2 \sum_{\tau_1 \text{ in triangulation of } \Omega_j''} \cdot \sum_{|\alpha|=2} \|D^\alpha(\psi\Theta_j)\|_{\psi^{-1},\tau_1}$$
$$\leq Ch^2\{\sum_{\tau_1}\sum_{|s+t|=2}(\|\psi D^s\Theta_j\|^2_{\psi^{-1},\tau_1} + \|\Theta_j D^t\psi\|_{\psi^{-1},\tau_1}\} \tag{4.16}$$

Analyzing the components in the sum on the R.H.S. of (4.16) individually: from Navert [12] quoted in section 2 as Lemma 2.1,

$$\sum_{\tau_1}\sum_{|s|=2}\|\Theta_j D^s\psi\|^2_{\psi^{-1},\tau_1} \leq Ch^{-2}\|\Theta_j\psi_{\underline{v}}\|^2_{\psi^{-1},\Omega_j''} \tag{4.17; a}$$
$$\leq C\gamma^{-1}h^{-3}\|\Theta_j\|_{-\psi_{\underline{v}},\Omega_j''} + Ch^s\|\Theta_j\|^2_{\psi}$$

From Lemma 2.1 also: $\|\Theta_{j,\underline{v}}\psi_{\underline{v}}\|_{\psi^{-1},\Omega_j''} \leq C(\gamma h)^{-2}\|\Theta_{j,\underline{v}}\|^2_{\psi} + Ch^s\|\Theta_j\|^2_{\psi}$

and from the inverse estimate for S^h and Lemma 2.1:

$$\|\Theta_{j,\cdot\underline{v}^{\perp}}\psi_{\underline{v}}\|_{\psi^{-1},\Omega_j''} \leq C\gamma^{-1}h^{-3}\|\Theta_j\|^2_{-\psi_{\underline{v}}} + Ch^s\|\Theta_j\|_{\psi}$$

Further, Lemma 2.1 gives,

$$\{\|\Theta_{j,\underline{v}^{\perp}}\psi_{\underline{v}}\|_{\psi^{-1},\Omega_j''} + \|\Theta_{j,\underline{v}}\psi_{\underline{v}^{\perp}}\|_{-\psi^{-1},\Omega_j''}\} \leq C\gamma^{-1}h^{-3}\|\Theta_j\|^2_{-\psi_{\underline{v}}} + Ch^s\|\Theta_j\|^2_{\psi}.$$

The previous three inequalities then give:

$$\sum_{\tau_j} \sum_{\substack{|s|=1 \\ |t|=1}} ||D^s\psi D^t\Theta_j||^2_{\psi^{-1}} \leq C\{(\gamma h)^{-2}||\Theta_{j,\underline{v}}||^2_\psi$$

$$+\gamma^{-1}h^{-3}||\Theta_j||^2_{-\psi_{\underline{v}}} + \gamma^{-1}h^{-3}||\Theta||^2_{-\psi_{\underline{v}}}\} + Ch^s||\Theta_j||^2_\psi \qquad (4.17;b)$$

From the higher order inverse estimate: $|\Theta_j|_{2,\tau_j} \leq Ch^{-1}||\nabla\Theta_j||_{\tau_j}$, $\Theta_j \in S^h$, we bound the last component of (4.16) as:

$$\sum_{\tau_j} \sum_{|\alpha|=2} ||\psi D^\alpha \Theta_j||^2_{\psi^{-1},\Omega_j''} \leq Ch^{-2}||\nabla\Theta_j||^2_\psi + Ch^2||\Theta_j||^2_\psi. \qquad (4.17;c)$$

Combining (4.17;a,b,c) into (4.16) gives

$$\begin{aligned} ||\nabla(\psi\Theta_j - I_h(\psi\Theta_j))||_{\psi^{-1}} &\leq Ch^2\{\gamma^{-1}h^{-3}||\Theta_j||^2_{-\psi_{\underline{v}}} \\ &+ (\gamma h)^{-2}||\Theta_{j,\underline{v}}||^2_\psi + h^{-2}||\nabla\Theta_j||^2_\psi\} + Ch^s||\Theta_j||^2_\psi. \end{aligned} \qquad (4.18)$$

Thus, as $\epsilon_0 = 0(h)$,

$$\begin{aligned} (\epsilon_0 - \epsilon)T_2 \leq (\epsilon_0 - \epsilon)C(\gamma)||\nabla(u^{j-1} - U^{j-1})||^2_\psi \\ + C\{\gamma^{-1}||\Theta_j||^2_{-\psi_{\underline{v}}} + (\epsilon_0 - \epsilon)\gamma^{-2}||\Theta_{j,\underline{v}}||^2_\psi \\ + \gamma^{-1}(\epsilon_0 - \epsilon)||\nabla\Theta_j||^2_\psi\} + Ch^s||\Theta_j||^2_\psi \end{aligned} \qquad (4.19)$$

Using (4.19) and

$$\begin{aligned} (\epsilon_0 - \epsilon)T_1 \leq (\epsilon_0 - \epsilon)C(\gamma)||\nabla(u^{j-1} - U^{j-1})||^2_\psi + (\epsilon_0 - \epsilon)\gamma^{-1}||\nabla\Theta_j||^2_\psi \\ + \gamma^{-1}||\Theta_j||^2_{-\psi_{\underline{v}}} + Ch^s||\Theta_j||^2_\psi \end{aligned}$$

(which follows from (4.15;b)) for γ sufficiently large in (4.12) gives:

$$\begin{aligned} |||\Theta_j|||^2_{\Omega_j'} &\leq C(|||\Theta_j|||^2_\psi + ||\Theta||^2_{-\psi_{\underline{v}}}) \leq \\ &\leq C\,\epsilon_0^s|||\Theta_j|||^2_{\Omega_j''} + C(\epsilon_0 - \epsilon)||\nabla(u^{j-1} - U^{j-1})||^2_\psi. \end{aligned} \qquad (4.20)$$

As $||w||_\psi \leq C||w||_{\Omega_j''}$ we have

$$|||\theta_j|||^2_{\Omega_j'} \leq C\epsilon_0^s|||\Theta_j|||^2_{\Omega_j''} + C(\epsilon_0 - \epsilon)||\nabla(u^{j-1} - U^{j-1})||^2_{\Omega_j''}. \qquad (4.21)$$

To bound the first term in the R.H.S.of above, note that

$$|||\Theta_j|||_{\Omega_j''} \le |||u^j - \mu_j(u^j)|||_{\Omega_j''} + |||u^j - U^j||_{\Omega_j''}. \tag{4.22}$$

From the global estimates for $u^j - Uj$ in Axelsson and Layton [3; Theorem 4.1] applied to Ω_j'' we have

$$|||u^j - U^j|||_{\Omega_j''} \le |||u^j - U^j|||_{\Omega} \le Ch^{k+1/2}(\sum_{\ell=1} ||u^\ell||_{k+1}) \tag{4.23}$$

Letting $\overline{g}_j$ be some $S^h(\Omega_j)$ function which interpolates u^j on $\partial\Omega_j'$ we then have, again from the global estimates in [3; Appendix 1] applied to Ω_j'' that $|||u^j - \mu_j(u^j)|||_{\Omega_j''} \le Ch^{k+1/2}|u^j - \overline{g}_j|_{k+1,\Omega_j''}$.

Using (4.24) (4.23) and (4.22) in (4.21) gives $|||\Theta_j|||^2_{\Omega_j} \le C(\epsilon_0 - \epsilon)$ $||\nabla(u^{j-1} - U^{j-1})||^2_{\Omega_j''}$.

Local error estimates for the artificial viscosity projection operator μ_j follow from, e.g. Navert [12; Corollary 4.3 p. 52]

$$\begin{aligned} |||u^j - \mu_j(u^j)|||_{\Omega_j'} \le C\{&h^{k+1/2}(|u^j|_{k+1,\Omega_j''} + |u^j|_{k+1,\Gamma_j''}) \\ &+ \epsilon_0^{s+1/2}(|u^j|_{1,\Omega_{k-1}'} + |u^j|_{1,\Gamma_{j-1}'}). \end{aligned} \tag{4.26}$$

Using the triangle inequality: $|||u^j - U^j|||_{\Omega_j'} \le |||\Theta_j|||_{\Omega_j'} + |||u^j - \mu_j(u^j)|||_{\Omega_j''}$, (4.25) and (4.26) we obtain abound for $|||u^j - U^j|||_{\Omega_j'}$:

$$\begin{aligned} |||u^j - U^j|||_{\Omega_j'} \le C|||u^{j-1} - U^{j-1}|||_{\Omega_j''} &+ Ch^{k+1/2}(|u^j|_{k+1,\Omega_j''} + |u^j|_{k+1,\Gamma_j''}) \\ + C\epsilon_0^s\{\sum_{\ell=1} |||u^\ell||| &+ h^{1/2}|\overline{g}_j|_{1,\Omega_j''} + |u^j|_{1,\Omega_{j-1}'} + |u^j|_{1,\Gamma_{j-1}'}\}. \end{aligned} \tag{4.27}$$

Letting $\overline{g}_j = I_h(u^j)$, which satisfies: $|\overline{g}_j|_{1,\Omega_j''} \le C||u^j||_{1,\Omega_j''}$ applying the trace theorem, using Corollary 2.1 and picking s sufficently high reduces (4.27) to:

$$\begin{aligned} |||u^j - U^j|||_{\Omega_j'} \le &C|||u^{j-1} - u^{j-1}|||_{\Omega_{j-1}'} \\ &+ Ch^{k+1/2}[\sum_{\ell=0}^{j=1}(\epsilon_0 - \epsilon)^\ell ||f||_{k+1+2\ell,\Omega_{j-\ell-1}'}\} \\ &+ C\epsilon_0^s[\sum_{\ell=0}^{j-1}(\epsilon_0 - \epsilon)^\ell ||f||_{2+2\ell,\Omega_{j-\ell--1}''}] \end{aligned}$$

The theorem now follows by induction.

5. Conclusions. It has been demonstrated that a defect-correction iteration method can improve the interior discretization by one order for every iteration up to an order, which is bounded by the polynomial degree of the finite element basis functions. This improvement is uniform in ϵ. The discretization errors in the L_2-norm are at most 1/2 order lower than the optimal order. To achieve this accuracy we must move slightly further inside Ω away from downstream and characteristic layers as we iterate. Another advantage of using defect-correction is that at every step we need to solve a linear algebraic system with a matrix which has a better eigenvalue distribution and can be better preconditioned by incomplete factorizations methods than the given high order approximation matrix. The eigenvalue distribution is favorable when the smallest real part of the eigenvalues is sufficently large and the corresponding imaginar parts are small. For the use of such preconditioning methods for a generalized conjugate gradient method to solve similar systems, see [15].

One major advantage of attaining this high accuracy in smooth region is computational economy. High resolution in smooth regions is achieved with few points there so gridpoints and computational effort in practical problems can be redistributed into layer areas, should information be sought there.

The work of the second author was partially supported by N.S.F. grant DMS 8701762.

References

[1] O. Axelsson, On the numerical solution of convection dominated, convection-diffusion problems, in: *Math. Meth. Energy Res.* (K.I. Gross, e.d.), SIAM, Philadelpha, 1984.

[2] O. Axelsson, Stability and error estimates of Galerkin finite element aproximations for convection-diffusion equations, I.M.A. Journal Numerical Analysis, 1 (1981), 329 -345.

[3] O. Axelsson and W. Layton, Defect correction methods for convection dominated, convection-diffusion problems, M.A.N. (formerly R.A.I.R.O. J. Numer. Anal.), 24 (1990), 423-455.

[4] W. Eckhaus, Boundary layers in linear elliptic singular perturbation problems, SIAM Review, 14 (1972), 225-270.

[5] V. Ervin and W. Layton, High resolution minimal storage algorithms for convection dominated, convection diffusion equations, pp. 1173-1201 in: *Trans. of the Fourth Army Conf.on Appl. Math. and Comp.*, 1987.

[6] V. Ervin and W. Layton, An analysis of a defect correction method for a model convection diffusion equation, SIAM J. Numer. Anal., 26 (1989), 169-179.

[7] P. W. Hemker, Mixed defect correction iteration for the accurate solution of the convecton diffusion equations, pp. 485-501 in: *Multigrid*

Methods, L.N.M. vol. 960, (W. Hackbusch and U. Trottenberg, eds.) Springer Verlag, Berlin 1982.

8] P. W. Hemker, The use of defect correction for the solution of a singularly perturbed o.d.e., preprint.

9] C. Johnson and U. Navert, An analysis of some finite element methods for advection diffusion problems, in: *Anal. and Numer. Approaches to Asym. Probs. in Analysis* (O. Axelsson, L. S. Frank and A. van der Sluis, eds.) North Holland, 1981, 99-116.

0] C. Johnson, U. Navert and J. Pitkaranta, Finite element methods for linear hyperbolic problems, Comp. Meth. Appl. Mech. Eng., 45 (1984), 285-312.

1] C. Johnson, A. H. Schatz and L. B. Wahlbin, Crosswind smear and pointwise errors in streamline diffusion finite element methods, Math. Comp., 49 (1987), 25-38.

2] U. Navert, A finite element method for convection diffusion problems, Thesis, Chalmers Inst. of Tech., 1982.

3] O. Axelsson and W. Layton, Optimal interior estimates for the defect-correction, finite element approach to 2-D convection-diffusion problems, ICMA report 88-116, Univ. of Pittsburgh, 1988.

4] V. Pereyra, Iterated defect corrections for nonlinear operator equations, Num. Math., 10 (1967) 316-323.

5] O. Axelsson, V. Eijkhout, B. Polman and P. Vassilevski, Incomplete block-matrix factorization iterative methods for convection-diffusion problems, BIT, to appear.

6] J. Boland and W. Layton, Error analysis of finite element methods for steady natural convection problems, to appear in: Numerical Functional Anal. and Opt.

7] P. W. Hemker and S. P. Spekreijse, Multigrid solution of the steady Euler equations, CWI report NM-R8507, (to appear in Appl. Num. Math.), 1985.

8] S. P. Spekreijse, Second order accurate upwind solutions of the 2-D steady Euler equations by use of a defect correction method, CWI Report NM-R8520, (to appear in: Proc. 2nd European Multigrid Conf.), 1985.

List of speakers

Robert Beauwens,
Fac. des Sciences Appl., Serv. de Métrol. Nucl., Univ. Libre de Bruxelles, Av. F.-D. Roosevelt 50, 1050 Bruxelles 5, Belgium

Peter Deuflhard,
Konrad-Zuse-Zentrum für Inform. Technik Berlin, Heilbronner Str. 10, D-1000 Berlin 31, Germany

Jocelyne Erhel,
IRISA, Campus de Banlieu, 35042 Rennes Cedex, France

Victor Eijkhout,
Faculty of Mathematics and Informatics, University of Nijmegen, Toernooiveld 5, 6525 ED Nijmegen, The Netherlands

Richard Ewing,
Director Inst. for Scientific Computation, University of Wyoming, Laramie, Wyoming 82081, USA

Gene Golub,
Department of Comp. Science, Stanford University, Stanford Calif 94305, USA

Danny Goovaerts,
Department of Computer Science, Celestijnenlaan 200A, B-3030 Leuven, Belgium

Ivar Gustafsson,
Informationsbehandlung, Chalmers University, S-41296 Göteborg, Sweden

Lennart Johnson,
Thinking Machines Corporation, 245 1st Street, Cambridge, Ma 02142, USA

Rik Kaasschieter,
TNO Institute of Appl. Geosci., P.O. Box 285, NL-2600 AG Delft, The Netherlands

Karl-Erik Karlsson,
Laboratory for Electrical Engeneering, Electrical Analysis Methods, ABB Corporate Research, Dept. KEB, S-72178 Västeras, Sweden

Mohammed Khalil,
Delft University of Technology, Fac. of Techn. Mathematics, and Informatics, P.O. Box 356, NL-2600 AJ Delft, The Netherlands

David Kincaid,
Center for Numer. Analysis, RLM 13.150, The University of Texas at Austin, Austin, TX 78712-1067, USA

Lily Kolotilina,
Steklov Institute of Math. Sciences, LOMI, Fontanka 27, Leningrad 191011, USSR

Arnold Krechel,
Gesellschaft für Mathematik und Datenverarbeitung mbH, Institut für Methodische Grundlagen, Postfach 1240, D-5205 St. Augustin 1, Germany

William Layton,
Mathematics Department, 301 Thackery Hall, University of Pittsburgh, Pittsburgh PA 15260, USA

Jos Maubach,
Faculty of Mathematics and Informatics, University of Nijmegen, Toernooiveld 5, 6525 ED Nijmegen, The Netherlands

Michael Navon,
Department of Mathematics, Florida State University, Tallahassee, FL 32306, USA

Yvan Notay,
Université Libre de Bruxelles, Service de Métrologie Nucl., Fac. des Sciences Appliq., Av F.-D. Roosevelt 50, B-1050 Bruxelles, Belgium

Ahmed Sameh,
University of Illinois, CSRD, 305 Talbot Laboratory, Urbana, IL 61801-2932, USA

Henk van der Vorst,
Delft University of Technology, Dept. of Mathematics and Informatics, P.O. Box 356, NL-2600 AJ Delft, The Netherlands

Rüdiger Weiss,
University Karlsruhe, Rechenzentrum, Zirkel 2, D-7500 Karlsruhe, Germany

David Young,
Center for Numerical Analysis, RLM 13.150, The University of Texas at Austin, Austin, TX 78713-8510, USA

Harry Yserentant,
Fachbereich Mathematik der Universität Dortmund, D-4600 Dortmund 50, Germany

Lecture Notes in Mathematics

Edited by A. Dold, B. Eckmann and F. Takens

Editorial Policy

for the publication of proceedings of conferences and other multi-author volumes

Lecture Notes aim to report new developments – quickly, informally and at a high level. The following describes criteria and procedures for multi-author volumes. For convenience we refer throughout to "proceedings" irrespective of whether the papers were presented at a meeting.

The editors of a volume are strongly advised to inform contributors about these points at an early stage.

§ 1. One (or more) expert participant(s) should act as the scientific editor(s) of the volume. They select the papers which are suitable (cf. §§ 2 – 5) for inclusion in the proceedings, and have them individually refereed (as for a journal). It should not be assumed that the published proceedings must reflect conference events in their entirety. The series editors will normally not interfere with the editing of a particular proceedings volume – except in fairly obvious cases, or on technical matters, such as described in §§ 2 – 5. The names of the scientific editors appear on the cover and title page of the volume.

§ 2. The proceedings should be reasonably homogeneous i.e. concerned with a limited and well-defined area. Papers that are essentially unrelated to this central topic should be excluded. One or two longer survey articles on recent developments in the field are often very useful additions. A detailed introduction on the subject of the congress is desirable.

§ 3. The final set of manuscripts should have at least 100 pages and preferably not exceed a total of 400 pages. Keeping the size below this bound should be achieved by stricter selection of articles and NOT by imposing an upper limit on the length of the individual papers.

§ 4. The contributions should be of a high mathematical standard and of current interest. Research articles should present new material and not duplicate other papers already published or due to be published. They should contain sufficient background and motivation and they should present proofs, or at least outlines of such, in sufficient detail to enable an expert to complete them. Thus summaries and mere announcements of papers appearing elsewhere cannot be included, although more detailed versions of, for instance, a highly technical contribution may well be published elsewhere later.
Contributions in numerical mathematics may be acceptable without formal theorems resp. proofs provided they present new algorithms solving problems (previously unsolved or less well solved) or develop innovative qualitative methods, not yet amenable to a more formal treatment.
Surveys, if included, should cover a sufficiently broad topic, and should normally not just review the author's own recent research. In the case of surveys, exceptionally, proofs of results may not be necessary.

§ 5. "Mathematical Reviews" and "Zentralblatt für Mathematik" recommend that papers in proceedings volumes carry an explicit statement that they are in final form and that no similar paper has been or is being submitted elsewhere, if these papers are to be considered for a review. Normally, papers that satisfy the criteria of the Lecture Notes in Mathematics series also satisfy this requirement, but we strongly recommend that each such paper carries the statement explicitly.

§ 6. Proceedings should appear soon after the related meeting. The publisher should therefore receive the complete manuscript (preferably in duplicate) including the Introduction and Table of Contents within nine months of the date of the meeting at the latest.

§ 7. Proposals for proceedings volumes should be sent to one of the editors of the series or to Springer-Verlag Heidelberg. They should give sufficient information on the conference, and on the proposed proceedings. In particular, they should include a list of the expected contributions with their prospective length. Abstracts or early versions (drafts) of the contributions are helpful.

General Remarks

Lecture Notes are printed by photo-offset from the master-copy delivered in camera-ready form by the authors of monographs, resp. editors of proceedings volumes. For this purpose Springer-Verlag provides technical instructions for the preparation of manuscripts. Volume editors are requested to distribute these to all contributing authors of proceedings volumes. Some homogeneity in the presentation of the contributions in a multi-author volume is desirable.
Careful preparation of manuscripts will help keep production time short and ensure a satisfactory appearance of the finished book. The actual production of a Lecture Notes volume normally takes approximately 8 weeks.
For monograph manuscripts typed or typeset according to our instructions, Springer-Verlag can, if necessary, contribute towards the preparation costs at a fixed rate.

Authors of monographs receive 50 free copies of their book. Editors of proceedings volumes similarly receive 50 copies of the book and are responsible for redistributing these to authors etc. at their discretion. No reprints of individual contributions can be supplied. No royalty is paid on Lecture Notes volumes.
Volume authors and editors are entitled to purchase further copies of their book for their personal use at a discount of 33.3 %, other Springer mathematics books at a discount of 20 % directly from Springer-Verlag. Authors contributing to proceedings volumes may purchase the volume in which their article appears at a discount of 20 %.

Commitment to publish is made by letter of intent rather than by signing a formal contract. Springer-Verlag secures the copyright for each volume.

LECTURE NOTES

ESSENTIALS FOR THE PREPARATION
OF CAMERA-READY MANUSCRIPTS

Springer
Springer-Verlag
Berlin Heidelberg New York
London Paris Tokyo Hong Kong

he preparation of manuscripts which are to be reproduced by photo-ffset require special care. Manuscripts which are submitted in tech-ically unsuitable form will be returned to the author for retyping. here is normally no possibility of carrying out further corrections fter a manuscript is given to production. Hence it is crucial that he following instructions be adhered to closely. If in doubt, please end us 1 - 2 sample pages for examination.

eneral. The characters must be uniformly black both within a single haracter and down the page. Original manuscripts are required: pho-ocopies are acceptable only if they are sharp and without smudges.

n request, Springer-Verlag will supply special paper with the text rea outlined. The standard TEXT AREA (OUTPUT SIZE if you are using a 4 point font) is 18 x 26.5 cm (7.5 x 11 inches). This will be scale-educed to 75% in the printing process. If you are using computer ypesetting, please see also the following page.

ake sure the TEXT AREA IS COMPLETELY FILLED. Set the margins so that hey precisely match the outline and type right from the top to the ottom line. (Note that the page number will lie outside this area). ines of text should not end more than three spaces inside or outside he right margin (see example on page 4).

ype on one side of the paper only.

pacing and Headings (Monographs). Use ONE-AND-A-HALF line spacing in he text. Please leave sufficient space for the title to stand out learly and do NOT use a new page for the beginning of subdivisons of hapters. Leave THREE LINES blank above and TWO below headings of uch subdivisions.

pacing and Headings (Proceedings). Use ONE-AND-A-HALF line spacing n the text. Do not use a new page for the beginning of subdivisons f a single paper. Leave THREE LINES blank above and TWO below hea-ings of such subdivisions. Make sure headings of equal importance re in the same form.

he first page of each contribution should be prepared in the same ay. The title should stand out clearly. We therefore recommend that he editor prepare a sample page and pass it on to the authors ogether with these instructions. Please take the following as an xample. Begin heading 2 cm below upper edge of text area.

MATHEMATICAL STRUCTURE IN QUANTUM FIELD THEORY

John E. Robert
Mathematisches Institut, Universität Heidelberg
Im Neuenheimer Feld 288, D-6900 Heidelberg

lease leave THREE LINES blank below heading and address of the uthor, then continue with the actual text on the same page.

ootnotes. These should preferable be avoided. If necessary, type hem in SINGLE LINE SPACING to finish exactly on the outline, and se-arate them from the preceding main text by a line.

Symbols. Anything which cannot be typed may be entered by hand i BLACK AND ONLY BLACK ink. (A fine-tipped rapidograph is suitable fo this purpose; a good black ball-point will do, but a pencil wil not). Do not draw straight lines by hand without a ruler (not even i fractions).

Literature References. These should be placed at the end of each pa per or chapter, or at the end of the work, as desired. Type them wit single line spacing and start each reference on a new line. Follo "Zentralblatt für Mathematik"/"Mathematical Reviews" for abbreviate titles of mathematical journals and "Bibliographic Guide for Editor and Authors (BGEA)" for chemical, biological, and physics journals Please ensure that all references are COMPLETE and ACCURATE.

IMPORTANT

Pagination. For typescript, <u>number pages in the upper right-hand cor ner in LIGHT BLUE OR GREEN PENCIL ONLY</u>. The printers will insert th final page numbers. For computer type, you may insert page number (1 cm above outer edge of text area).

It is safer to number pages AFTER the text has been typed and correc ted. Page 1 (Arabic) should be THE FIRST PAGE OF THE ACTUAL TEXT. Th Roman pagination (table of contents, preface, abstract, acknowledge ments, brief introductions, etc.) will be done by Springer-Verlag.

If including running heads, these should be aligned with the insid edge of the text area while the page number is aligned with the out side edge noting that <u>right</u>-hand pages are <u>odd</u>-numbered. Runnin heads and page numbers appear on the same line. Normally, the runnin head on the left-hand page is the chapter heading and that on th right-hand page is the section heading. Running heads should <u>not</u> b included in proceedings contributions unless this is being done con sistently by all authors.

Corrections. When corrections have to be made, cut the new text t fit and paste it over the old. White correction fluid may also b used.

Never make corrections or insertions in the text by hand.

If the typescript has to be marked for any reason, e.g. for provisio nal page numbers or to mark corrections for the typist, this can b done VERY FAINTLY with BLUE or GREEN PENCIL but NO OTHER COLOR: thes colors do not appear after reproduction.

COMPUTER-TYPESETTING. Further, to the above instructions, please not with respect to your printout that
- the characters should be sharp and sufficiently black;
- it is not strictly necessary to use Springer's special typin paper. Any white paper of reasonable quality is acceptable.

If you are using a significantly different font size, you shoul modify the output size correspondingly, keeping length to breadt ratio 1 : 0.68, so that scaling down to 10 point font size, yields text area of 13.5 x 20 cm (5 3/8 x 8 in), e.g.

Differential equations.: use output size 13.5 x 20 cm.

Differential equations.: use output size 16 x 23.5 cm.

Differential equations.: use output size 18 x 26.5 cm.

Interline spacing: 5.5 mm base-to-base for 14 point characters (stan dard format of 18 x 26.5 cm).
If in any doubt, please send us 1 - 2 sample pages for examination We will be glad to give advice.

Vol. 1290: G. Wüstholz (Ed.), Diophantine Approximation and Transcendence Theory. Seminar, 1985. V, 243 pages. 1987.

Vol. 1291: C. Mœglin, M.-F. Vignéras, J.-L. Waldspurger, Correspondances de Howe sur un Corps p-adique. VII, 163 pages. 1987

Vol. 1292: J.T. Baldwin (Ed.), Classification Theory. Proceedings, 1985. VI, 500 pages. 1987.

Vol. 1293: W. Ebeling, The Monodromy Groups of Isolated Singularities of Complete Intersections. XIV, 153 pages. 1987.

Vol. 1294: M. Queffélec, Substitution Dynamical Systems – Spectral Analysis. XIII, 240 pages. 1987.

Vol. 1295: P. Lelong, P. Dolbeault, H. Skoda (Réd.), Séminaire d'Analyse P. Lelong – P. Dolbeault – H. Skoda. Seminar, 1985/1986. VII, 283 pages. 1987.

Vol. 1296: M.-P. Malliavin (Ed.), Séminaire d'Algèbre Paul Dubreil et Marie-Paule Malliavin. Proceedings, 1986. IV, 324 pages. 1987.

Vol. 1297: Zhu Y.-l., Guo B.-y. (Eds.), Numerical Methods for Partial Differential Equations. Proceedings. XI, 244 pages. 1987.

Vol. 1298: J. Aguadé, R. Kane (Eds.), Algebraic Topology, Barcelona 1986. Proceedings. X, 255 pages. 1987.

Vol. 1299: S. Watanabe, Yu.V. Prokhorov (Eds.), Probability Theory and Mathematical Statistics. Proceedings, 1986. VIII, 589 pages. 1988.

Vol. 1300: G.B. Seligman, Constructions of Lie Algebras and their Modules. VI, 190 pages. 1988.

Vol. 1301: N. Schappacher, Periods of Hecke Characters. XV, 160 pages. 1988.

Vol. 1302: M. Cwikel, J. Peetre, Y. Sagher, H. Wallin (Eds.), Function Spaces and Applications. Proceedings, 1986. VI, 445 pages. 1988.

Vol. 1303: L. Accardi, W. von Waldenfels (Eds.), Quantum Probability and Applications III. Proceedings, 1987. VI, 373 pages. 1988.

Vol. 1304: F.Q. Gouvêa, Arithmetic of p-adic Modular Forms. VIII, 121 pages. 1988.

Vol. 1305: D.S. Lubinsky, E.B. Saff, Strong Asymptotics for Extremal Polynomials Associated with Weights on ℝ. VII, 153 pages. 1988.

Vol. 1306: S.S. Chern (Ed.), Partial Differential Equations. Proceedings, 1986. VI, 294 pages. 1988.

Vol. 1307: T. Murai, A Real Variable Method for the Cauchy Transform, and Analytic Capacity. VIII, 133 pages. 1988.

Vol. 1308: P. Imkeller, Two-Parameter Martingales and Their Quadratic Variation. IV, 177 pages. 1988.

Vol. 1309: B. Fiedler, Global Bifurcation of Periodic Solutions with Symmetry. VIII, 144 pages. 1988.

Vol. 1310: O.A. Laudal, G. Pfister, Local Moduli and Singularities. V, 117 pages. 1988.

Vol. 1311: A. Holme, R. Speiser (Eds.), Algebraic Geometry, Sundance 1986. Proceedings. VI, 320 pages. 1988.

Vol. 1312: N.A. Shirokov, Analytic Functions Smooth up to the Boundary. III, 213 pages. 1988.

Vol. 1313: F. Colonius, Optimal Periodic Control. VI, 177 pages. 1988.

Vol. 1314: A. Futaki, Kähler-Einstein Metrics and Integral Invariants. IV, 140 pages. 1988.

Vol. 1315: R.A. McCoy, I. Ntantu, Topological Properties of Spaces of Continuous Functions. IV, 124 pages. 1988.

Vol. 1316: H. Korezlioglu, A.S. Ustunel (Eds.), Stochastic Analysis and Related Topics. Proceedings, 1986. V, 371 pages. 1988.

Vol. 1317: J. Lindenstrauss, V.D. Milman (Eds.), Geometric Aspects of Functional Analysis. Seminar, 1986–87. VII, 289 pages. 1988.

Vol. 1318: Y. Felix (Ed.), Algebraic Topology – Rational Homotopy. Proceedings, 1986. VIII, 245 pages. 1988

Vol. 1319: M. Vuorinen, Conformal Geometry and Quasiregular Mappings. XIX, 209 pages. 1988.

Vol. 1320: H. Jürgensen, G. Lallement, H.J. Weinert (Eds.), Semigroups, Theory and Applications. Proceedings, 1986. X, 416 pages. 1988.

Vol. 1321: J. Azéma, P.A. Meyer, M. Yor (Eds.), Séminaire de Probabilités XXII. Proceedings. IV, 600 pages. 1988.

Vol. 1322: M. Métivier, S. Watanabe (Eds.), Stochastic Analysis. Proceedings, 1987. VII, 197 pages. 1988.

Vol. 1323: D.R. Anderson, H.J. Munkholm, Boundedly Controlled Topology. XII, 309 pages. 1988.

Vol. 1324: F. Cardoso, D.G. de Figueiredo, R. Iório, O. Lopes (Eds.), Partial Differential Equations. Proceedings, 1986. VIII, 433 pages. 1988.

Vol. 1325: A. Truman, I.M. Davies (Eds.), Stochastic Mechanics and Stochastic Processes. Proceedings, 1986. V, 220 pages. 1988.

Vol. 1326: P.S. Landweber (Ed.), Elliptic Curves and Modular Forms in Algebraic Topology. Proceedings, 1986. V, 224 pages. 1988.

Vol. 1327: W. Bruns, U. Vetter, Determinantal Rings. VII,236 pages. 1988.

Vol. 1328: J.L. Bueso, P. Jara, B. Torrecillas (Eds.), Ring Theory. Proceedings, 1986. IX, 331 pages. 1988.

Vol. 1329: M. Alfaro, J.S. Dehesa, F.J. Marcellan, J.L. Rubio de Francia, J. Vinuesa (Eds.): Orthogonal Polynomials and their Applications. Proceedings, 1986. XV, 334 pages. 1988.

Vol. 1330: A. Ambrosetti, F. Gori, R. Lucchetti (Eds.), Mathematical Economics. Montecatini Terme 1986. Seminar. VII, 137 pages. 1988.

Vol. 1331: R. Bamón, R. Labarca, J. Palis Jr. (Eds.), Dynamical Systems, Valparaiso 1986. Proceedings. VI, 250 pages. 1988.

Vol. 1332: E. Odell, H. Rosenthal (Eds.), Functional Analysis. Proceedings, 1986–87. V, 202 pages. 1988.

Vol. 1333: A.S. Kechris, D.A. Martin, J.R. Steel (Eds.), Cabal Seminar 81–85. Proceedings, 1981–85. V, 224 pages. 1988.

Vol. 1334: Yu.G. Borisovich, Yu. E. Gliklikh (Eds.), Global Analysis – Studies and Applications III. V, 331 pages. 1988.

Vol. 1335: F. Guillén, V. Navarro Aznar, P. Pascual-Gainza, F. Puerta, Hyperrésolutions cubiques et descente cohomologique. XII, 192 pages. 1988.

Vol. 1336: B. Helffer, Semi-Classical Analysis for the Schrödinger Operator and Applications. V, 107 pages. 1988.

Vol. 1337: E. Sernesi (Ed.), Theory of Moduli. Seminar, 1985. VIII, 232 pages. 1988.

Vol. 1338: A.B. Mingarelli, S.G. Halvorsen, Non-Oscillation Domains of Differential Equations with Two Parameters. XI, 109 pages. 1988.

Vol. 1339: T. Sunada (Ed.), Geometry and Analysis of Manifolds. Procedings, 1987. IX, 277 pages. 1988.

Vol. 1340: S. Hildebrandt, D.S. Kinderlehrer, M. Miranda (Eds.), Calculus of Variations and Partial Differential Equations. Proceedings, 1986. IX, 301 pages. 1988.

Vol. 1341: M. Dauge, Elliptic Boundary Value Problems on Corner Domains. VIII, 259 pages. 1988.

Vol. 1342: J.C. Alexander (Ed.), Dynamical Systems. Proceedings, 1986–87. VIII, 726 pages. 1988.

Vol. 1343: H. Ulrich, Fixed Point Theory of Parametrized Equivariant Maps. VII, 147 pages. 1988.

Vol. 1344: J. Král, J. Lukeš, J. Netuka, J. Veselý (Eds.), Potential Theory – Surveys and Problems. Proceedings, 1987. VIII, 271 pages. 1988.

Vol. 1345: X. Gomez-Mont, J. Seade, A. Verjovski (Eds.), Holomorphic Dynamics. Proceedings, 1986. VII, 321 pages. 1988.

Vol. 1346: O. Ya. Viro (Ed.), Topology and Geometry – Rohlin Seminar. XI, 581 pages. 1988.

Vol. 1347: C. Preston, Iterates of Piecewise Monotone Mappings on an Interval. V, 166 pages. 1988.

Vol. 1348: F. Borceux (Ed.), Categorical Algebra and its Applications. Proceedings, 1987. VIII, 375 pages. 1988.

Vol. 1349: E. Novak, Deterministic and Stochastic Error Bounds in Numerical Analysis. V, 113 pages. 1988.

Vol. 1350: U. Koschorke (Ed.), Differential Topology. Proceedings, 1987. VI, 269 pages. 1988.

Vol. 1351: I. Laine, S. Rickman, T. Sorvali, (Eds.), Complex Analysis, Joensuu 1987. Proceedings. XV, 378 pages. 1988.

Vol. 1352: L.L. Avramov, K.B. Tchakerian (Eds.), Algebra – Some Current Trends. Proceedings, 1986. IX, 240 Seiten. 1988.

Vol. 1353: R.S. Palais, Ch.-l. Terng, Critical Point Theory and Submanifold Geometry. X, 272 pages. 1988.

Vol. 1354: A. Gómez, F. Guerra, M.A. Jiménez, G. López (Eds.), Approximation and Optimization. Proceedings, 1987. VI, 280 pages. 1988.

Vol. 1355: J. Bokowski, B. Sturmfels, Computational Synthetic Geometry. V, 168 pages. 1989.

Vol. 1356: H. Volkmer, Multiparameter Eigenvalue Problems and Expansion Theorems. VI, 157 pages. 1988.

Vol. 1357: S. Hildebrandt, R. Leis (Eds.), Partial Differential Equations and Calculus of Variations. VI, 423 pages. 1988.

Vol. 1358: D. Mumford, The Red Book of Varieties and Schemes. V, 309 pages. 1988.

Vol. 1359: P. Eymard, J.-P. Pier (Eds.), Harmonic Analysis. Proceedings, 1987. VIII, 287 pages. 1988.

Vol. 1360: G. Anderson, C. Greengard (Eds.), Vortex Methods. Proceedings, 1987. V, 141 pages. 1988.

Vol. 1361: T. tom Dieck (Ed.), Algebraic Topology and Transformation Groups. Proceedings, 1987. VI, 298 pages. 1988.

Vol. 1362: P. Diaconis, D. Elworthy, H. Föllmer, E. Nelson, G.C. Papanicolaou, S.R.S. Varadhan. École d'Été de Probabilités de Saint-Flour XV–XVII, 1985–87. Editor: P.L. Hennequin. V, 459 pages. 1988.

Vol. 1363: P.G. Casazza, T.J. Shura. Tsirelson's Space. VIII, 204 pages. 1988.

Vol. 1364: R.R. Phelps, Convex Functions, Monotone Operators and Differentiability. IX, 115 pages. 1989.

Vol. 1365: M. Giaquinta (Ed.), Topics in Calculus of Variations. Seminar, 1987. X, 196 pages. 1989.

Vol. 1366: N. Levitt, Grassmannians and Gauss Maps in PL-Topology. V, 203 pages. 1989.

Vol. 1367: M. Knebusch, Weakly Semialgebraic Spaces. XX, 376 pages. 1989.

Vol. 1368: R. Hübl, Traces of Differential Forms and Hochschild Homology. III, 111 pages. 1989.

Vol. 1369: B. Jiang, Ch.-K. Peng, Z. Hou (Eds.), Differential Geometry and Topology. Proceedings, 1986–87. VI, 366 pages. 1989.

Vol. 1370: G. Carlsson, R.L. Cohen, H.R. Miller, D.C. Ravenel (Eds.), Algebraic Topology. Proceedings, 1986. IX, 456 pages. 1989.

Vol. 1371: S. Glaz, Commutative Coherent Rings. XI, 347 pages. 1989.

Vol. 1372: J. Azéma, P.A. Meyer, M. Yor (Eds.), Séminaire de Probabilités XXIII. Proceedings. IV, 583 pages. 1989.

Vol. 1373: G. Benkart, J.M. Osborn (Eds.), Lie Algebras, Madison 1987. Proceedings. V, 145 pages. 1989.

Vol. 1374: R.C. Kirby, The Topology of 4-Manifolds. VI, 108 pages. 1989.

Vol. 1375: K. Kawakubo (Ed.), Transformation Groups. Proceedings, 1987. VIII, 394 pages, 1989.

Vol. 1376: J. Lindenstrauss, V.D. Milman (Eds.), Geometric Aspects of Functional Analysis. Seminar (GAFA) 1987–88. VII, 288 pages. 1989.

Vol. 1377: J.F. Pierce, Singularity Theory, Rod Theory, and Symmetry-Breaking Loads. IV, 177 pages. 1989.

Vol. 1378: R.S. Rumely, Capacity Theory on Algebraic Curves. III, 437 pages. 1989.

Vol. 1379: H. Heyer (Ed.), Probability Measures on Groups IX. Proceedings, 1988. VIII, 437 pages. 1989

Vol. 1380: H.P. Schlickewei, E. Wirsing (Eds.), Number Theory, Ul 1987. Proceedings. V, 266 pages. 1989.

Vol. 1381: J.-O. Strömberg, A. Torchinsky. Weighted Hardy Space. V, 193 pages. 1989.

Vol. 1382: H. Reiter, Metaplectic Groups and Segal Algebras. XI, 12 pages. 1989.

Vol. 1383: D.V. Chudnovsky, G.V. Chudnovsky, H. Cohn, M.B. Nathanso (Eds.), Number Theory, New York 1985–88. Seminar. V, 256 pages. 1989.

Vol. 1384: J. Garcia-Cuerva (Ed.), Harmonic Analysis and Parti Differential Equations. Proceedings, 1987. VII, 213 pages. 1989.

Vol. 1385: A.M. Anile, Y. Choquet-Bruhat (Eds.), Relativistic Fluid Dyn mics. Seminar, 1987. V, 308 pages. 1989.

Vol. 1386: A. Bellen, C.W. Gear, E. Russo (Eds.), Numerical Methods fc Ordinary Differential Equations. Proceedings, 1987. VII, 136 pages. 1989.

Vol. 1387: M. Petković, Iterative Methods for Simultaneous Inclusion c Polynomial Zeros. X, 263 pages. 1989.

Vol. 1388: J. Shinoda, T.A. Slaman, T. Tugué (Eds.), Mathematical Logi and Applications. Proceedings, 1987. V, 223 pages. 1989.

Vol. 1000: Second Edition. H. Hopf, Differential Geometry in the Large. V 184 pages. 1989.

Vol. 1389: E. Ballico, C. Ciliberto (Eds.), Algebraic Curves and Projectiv Geometry. Proceedings, 1988. V, 288 pages. 1989.

Vol. 1390: G. Da Prato, L. Tubaro (Eds.), Stochastic Partial Differenti Equations and Applications II. Proceedings, 1988. VI, 258 pages. 1989.

Vol. 1391: S. Cambanis, A. Weron (Eds.), Probability Theory on Vecto Spaces IV. Proceedings, 1987. VIII, 424 pages. 1989.

Vol. 1392: R. Silhol, Real Algebraic Surfaces. X, 215 pages. 1989.

Vol. 1393: N. Bouleau, D. Feyel, F. Hirsch, G. Mokobodzki (Eds. Séminaire de Théorie du Potentiel Paris, No. 9. Proceedings. VI, 26 pages. 1989.

Vol. 1394: T.L. Gill, W.W. Zachary (Eds.), Nonlinear Semigroups, Parti Differential Equations and Attractors. Proceedings, 1987. IX, 233 page 1989.

Vol. 1395: K. Alladi (Ed.), Number Theory, Madras 1987. Proceedings. V 234 pages. 1989.

Vol. 1396: L. Accardi, W. von Waldenfels (Eds.), Quantum Probability an Applications IV. Proceedings, 1987. VI, 355 pages. 1989.

Vol. 1397: P.R. Turner (Ed.), Numerical Analysis and Parallel Processing Seminar, 1987. VI, 264 pages. 1989.

Vol. 1398: A.C. Kim, B.H. Neumann (Eds.), Groups – Korea 1988. Pro ceedings. V, 189 pages. 1989.

Vol. 1399: W.-P. Barth, H. Lange (Eds.), Arithmetic of Complex Manifolds Proceedings, 1988. V, 171 pages. 1989.

Vol. 1400: U. Jannsen. Mixed Motives and Algebraic K-Theory. XIII, 24 pages. 1990.

Vol. 1401: J. Steprāns, S. Watson (Eds.), Set Theory and its Applications Proceedings, 1987. V, 227 pages. 1989.

Vol. 1402: C. Carasso, P. Charrier, B. Hanouzet, J.-L. Joly (Eds. Nonlinear Hyperbolic Problems. Proceedings, 1988. V, 249 pages. 1989

Vol. 1403: B. Simeone (Ed.), Combinatorial Optimization. Seminar, 198 V, 314 pages. 1989.

Vol. 1404: M.-P. Malliavin (Ed.), Séminaire d'Algèbre Paul Dubreil et Mari Paul Malliavin. Proceedings, 1987 – 1988. IV, 410 pages. 1989.

Vol. 1405: S. Dolecki (Ed.), Optimization. Proceedings, 1988. V, 22 pages. 1989.

Vol. 1406: L. Jacobsen (Ed.), Analytic Theory of Continued Fractions II Proceedings, 1988. VI, 142 pages. 1989.

Vol. 1407: W. Pohlers, Proof Theory. VI, 213 pages. 1989.

Vol. 1408: W. Lück, Transformation Groups and Algebraic K-Theory. XI 443 pages. 1989.

Vol. 1409: E. Hairer, Ch. Lubich, M. Roche. The Numerical Solution o Differential-Algebraic Systems by Runge-Kutta Methods. VII, 139 page 1989.

Vol. 1410: F.J. Carreras, O. Gil-Medrano, A.M. Naveira (Eds.), Differenti Geometry. Proceedings, 1988. V, 308 pages. 1989.